Managing [illegible]ration at Work

LEEDS COLLEGE OF BUILDING
WITHDRAWN FROM STOCK

Managing Noise and Vibration at Work

A practical guide to assessment, measurement and control

Tim South

ELSEVIER
BUTTERWORTH HEINEMANN

AMSTERDAM · BOSTON · HEIDELBERG · LONDON · NEW YORK · OXFORD
PARIS · SAN DIEGO · SAN FRANCISCO · SINGAPORE · SYDNEY · TOKYO

Elsevier Butterworth-Heinemann
Linacre House, Jordan Hill, Oxford OX2 8DP
30 Corporate Drive, Burlington, MA 01803

First published 2004

British Library Cataloguing in Publication Data
A catalogue record for this book is available from the British Library

Library of Congress Cataloguing in Publication Data
A catalogue record for this book is available from the Library of Congress

ISBN 0 7506 6342 1

For information on all Elsevier Butterworth-Heinemann publications visit our website at http://books.elsevier.com

Printed and bound in Great Britain by Biddles Ltd, King's Lynn, Norfolk

Contents

Preface

The European Union has issued two physical agents directives in the space of less than a year – Vibration in July 2002 and Noise in February 2003. In the UK we previously had no legal limits on vibration exposure, and a great many employees are exposed to noise levels falling below the first action level of the old directive, but above the new directive's lower exposure action value. Are we on the road to being over-regulated, burdening employers with unnecessary expense and making it virtually impossible for some of them to operate at all?

Large-scale programmes to measure noise or vibration can certainly be expensive, as can many actions intended to reduce exposure to these hazards. The new legislation will not automatically impose these extra costs on employers who are already managing these hazards effectively, but it does make it more important that those responsible for health and safety have a good understanding of how they arise, how exposure can be measured (and when it needs to be) and what sort of control measures are likely to work.

For the last few years I have been teaching courses for the Institute of Acoustics' Certificate of Competence in Workplace Noise Assessment, and more recently also for their Certificate in the Management of Occupational Exposure to Hand–arm Vibration. The participants in these courses come from various backgrounds – while many are health and safety managers, others work in the consultancy sector or are involved in health and safety issues as department managers. However, I find that the content of the courses is appropriate for all these participants. Better measurements are made by those who understand how the results will be used and better interpretations result from a knowledge of some of the practicalities of measurement. This book covers the syllabus of both these courses, which in turn were based on the Health and Safety Executive's view of what should be included. By including the two hazards in a single book I have, I hope, been able to point out some of the similarities of approach when dealing with the two hazards. Equally, I have been able to draw attention to some of the clear differences in the ways they must be managed.

Whole body vibration was more of a problem. Neither of the IOA courses addresses it, and I do not know of any others which do so. The subject is extremely complex, and to cover it even to the same depth as noise and hand–arm vibration would have taken a great many more pages than I thought were justified. At the same time, it is included in the Physical Agents (Vibration) Directive and I know from experience that many health and safety managers would like some guidance on how to manage the risks effectively in their organizations. Fortunately, in many workplaces the risks can be managed adequately without delving too deeply into the subject. I have tried to include enough guidance to assist those responsible for relatively low-level exposures to whole body vibration (perhaps to those driving delivery vehicles and fork-lift trucks). In these cases most health and safety professionals – particularly those with experience of noise and hand–arm vibration assessment – should be able to carry out a risk assessment without too much difficulty, and the WBV content of this book is geared towards helping them to do this. Those with responsibility for higher levels of WBV exposure, or with greater difficulties in assessing the risks, would be well advised to obtain specialist advice.

At the time of writing (early 2004) the wording of the UK regulations to implement both directives has not yet been finalized. The most significant issue to be resolved is whether the UK adopts exposure action and limit values for whole body vibration based on measurement of VDV or of A(8) (or on some combination). Both methods are covered in the relevant chapters.

How much mathematics to include in the book was an issue that caused much thought and discussion. Some potential readers will be deterred by the presence of any maths at all. Many others will see the maths as being the hardest part of the book to understand. If the book is to be of any use to those studying for the two certificate courses mentioned (and other similar ones) it has to include the mathematical procedures that are used on those courses. It is almost impossible, in any case, to assess workplace noise or vibration exposure without doing at least some calculations. On our courses, a great deal of time is spent in making sure that every participant can carry out a few basic calculations with the surprising variety of calculators that they bring along. This sort of help cannot be given in a book. Nevertheless, I have tried to make things easier for those who find maths a problem. Each type of calculation is accompanied by a worked example. Most of these examples (and some other mathematical material) are segregated from the rest of the text. This is not so it can be ignored completely, but to make it easier to read through smoothly on the first occasion and to return to spend more time following through the calculations later. Many chapters contain no maths at all, but I have included two (Chapters 5 and 11) which bring together the various types of calculation required when assessing noise and hand–arm vibration exposure, respectively.

I should issue one warning. Reading this book (or any other book) will not, on its own, make anyone competent to carry out assessments of exposure to either noise or vibration. Normally this will require some combination of formal study plus experience gained by working with those who are skilled in these fields.

I would recommend the courses accredited by the Institute of Acoustics (and similar courses operated by some other organizations) as an ideal preparation to take on responsibilities in this field. I hope, though, that this book will be useful to those who wish to consolidate and update their existing knowledge, as well as to those who are studying these subjects for the first time.

Tim South

Acknowledgements

I would like to extend my thanks to a number of people and organizations who have helped me to complete this book.

Terry Collins produced a number of the line drawings. Jonathon Crosby gave me a great deal of advice and help with the photographs. Others who helped with the photographs include Kevin Smith, Peter Griffiths, Mark Walter and Emily South. Alex Garry allowed me to use some of her measurement results. Jane Arnaud read and advised on some of the chapters. Thank you also to Mono Pumps Ltd, Spooner Industries Ltd and Leeds Metropolitan University for allowing me to use photographs taken on their premises.

I owe a debt to many other colleagues at Leeds Met, in the Institute of Acoustics and in various other companies and organizations with whom I have worked over the years on workplace noise and vibration matters. Many of their experiences and insights have found their way into these pages.

Finally, thank you to all those who have studied workplace noise and vibration on our courses at Leeds Metropolitan University during the last few years. It is during these courses that the need for a book on these topics became apparent.

None of those mentioned are to blame for any errors or omissions, for which I take full responsibility.

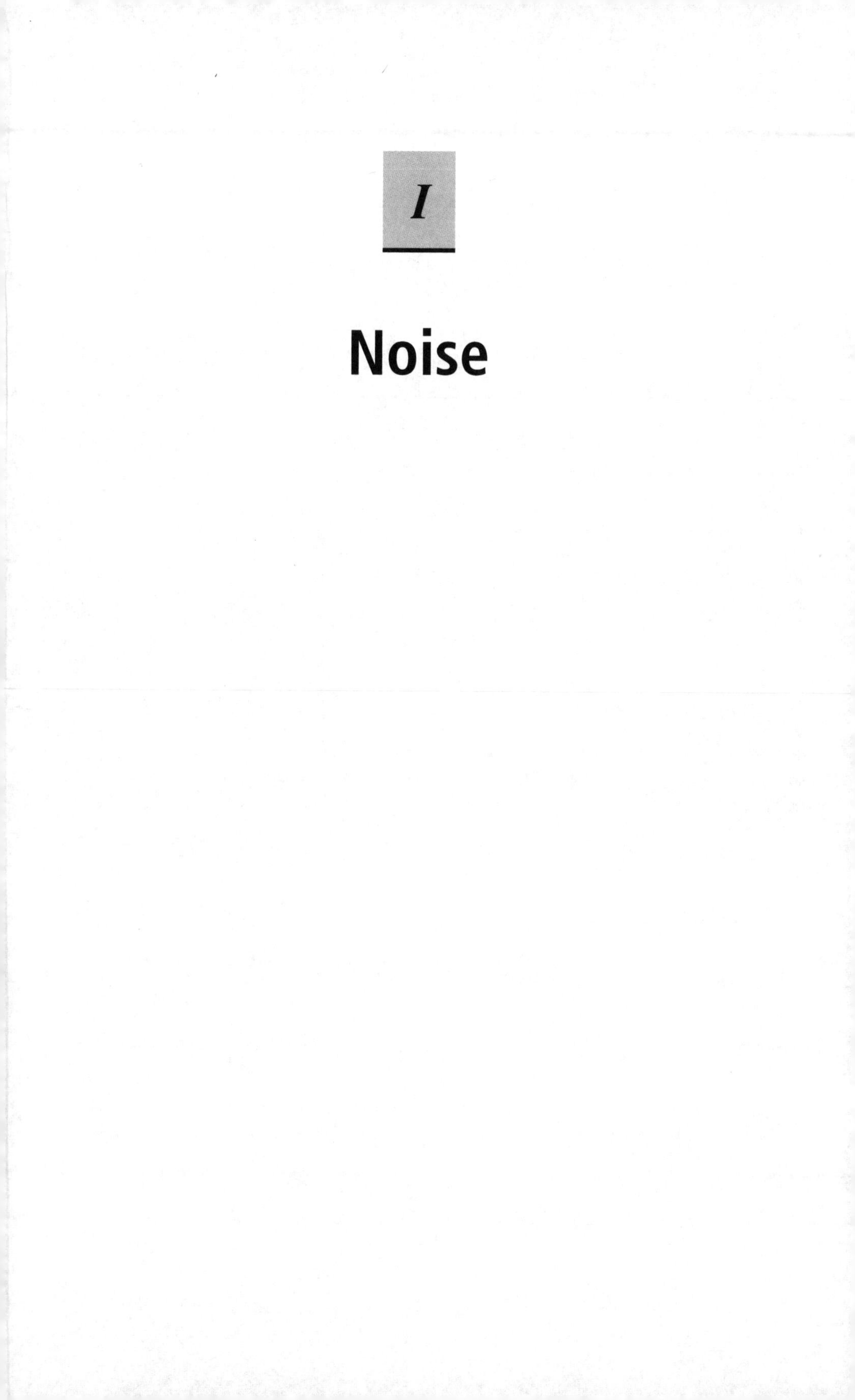

I

Noise

Noise and how it behaves

What is noise?

Sound arises when fluctuations in air pressure give rise to pressure waves which travel through the atmosphere. As they travel they will interact in various ways with their surroundings. Noise is a word which is normally applied to unwanted sound, and the sound present in most work situations is unwanted, so we normally talk about exposure to workplace noise rather than to workplace sound.

Mathematical descriptions of how sound behaves as it interacts with solid objects can be very complicated. Fortunately it is possible to produce full descriptions of the behaviour of sound in simple, idealized situations, and to use these to build up to more realistic situations.

The simplest type of sound wave would be a pure tone – a sine wave – moving in one direction without spreading out as it moves away from the source. If you could take a snapshot of the pressure variations along the direction it was moving in, you would get a picture such as the one in Figure 1.1.

This is rather difficult to draw, so it is normally easier to show the pressure variations as a graph of pressure against distance. It should be remembered when this is done, though, that sound is a longitudinal wave. In other words the air movements as the wave progresses are backwards and forwards in the same direction as the wave as a whole is travelling. This is different from a transverse wave such as a series of ripples on a water surface, in which case the water is moving up and down while the wave travels horizontally (Figure 1.2). If you could stand at one point as the sound wave travelled towards you, and plot the pressure as a function of time, you would also get a sinusoidal shape (Figure 1.3). This is assuming you could work very fast; the pressure might be varying up and down several thousand times a second!

Even for quite loud sounds, the actual pressure change is rather small. A 1 per cent fluctuation in atmospheric pressure would be associated with an intolerably

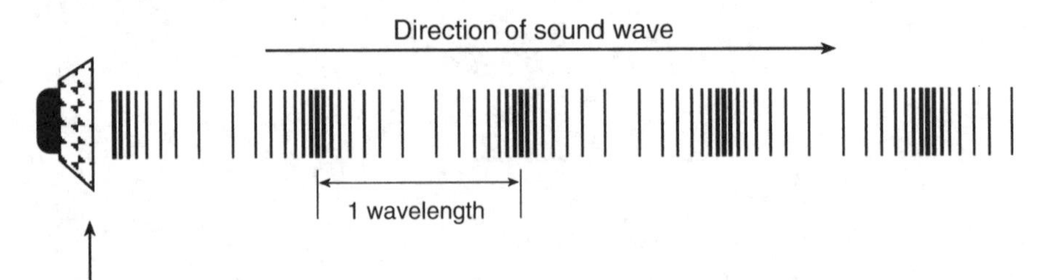

Figure 1.1 A representation of a sound wave.

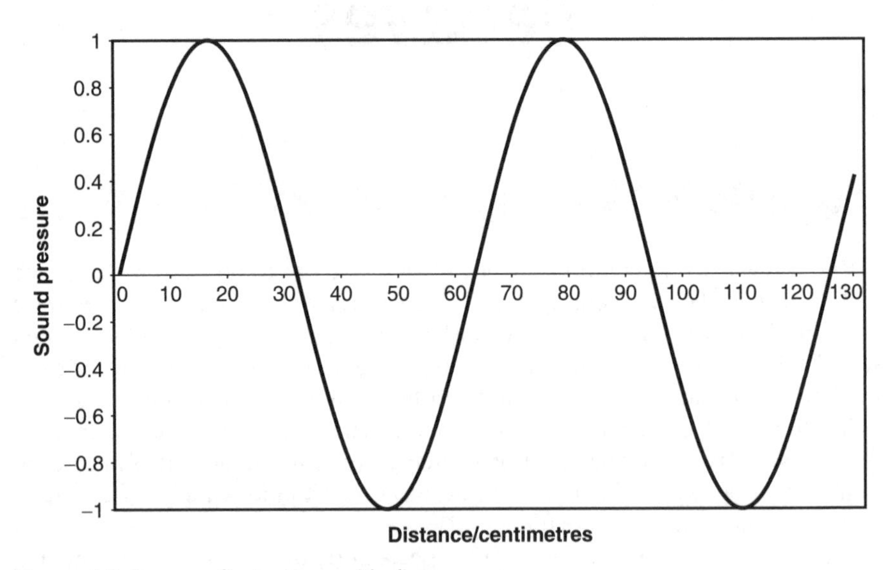

Figure 1.2 Pressure fluctuations with distance.

loud sound. By comparison, the atmospheric pressure can easily change by 2–3 per cent in the course of an ordinary week's weather. When we measure the magnitude of a sound wave, therefore, we concentrate on the deviations from ambient air pressure, and this deviation is normally called the sound pressure.

Pure tones, sine waves, frequency and wavelength

The sound wave described above will have a particular frequency. Essentially, the frequency of a wave is the number of complete waves which pass any particular point in the course of one second. The unit of frequency is the

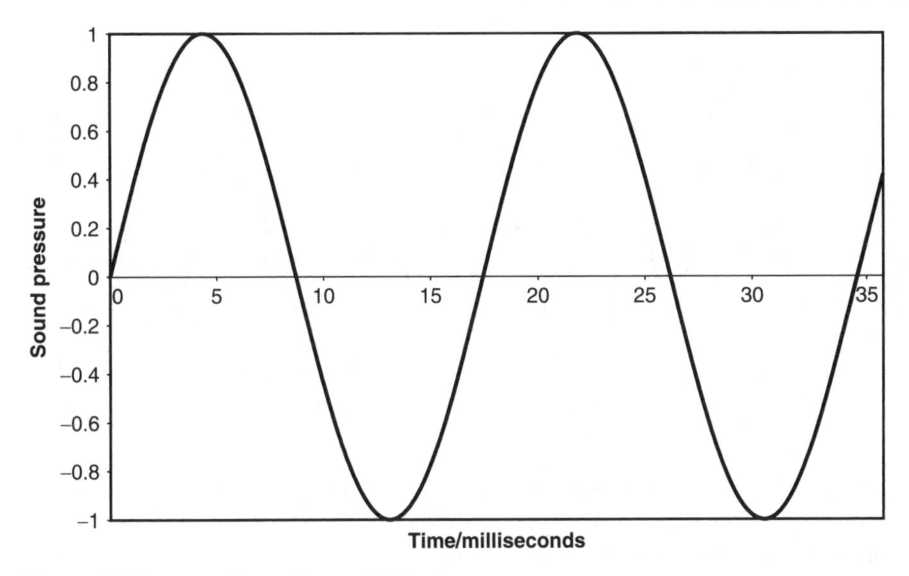

Figure 1.3 Pressure fluctuations with time.

Hertz. If the frequency is 1000 Hz (i.e. 1 kHz), then 1000 complete waves will arrive in one second. The frequency of a pure tone such as this is very closely related to its apparent pitch; a high pitched sound has a high frequency, while a low pitched tone will have a low frequency. The range of tones which are normally considered to be audible to human beings ranges from 20 to 20 000 Hz (or 20 kHz). This conceals a great deal of variation in the hearing abilities of different individuals. In particular, as human beings age they lose their ability to hear high frequencies. In any case, there is no sharp cut-off point at either end of the frequency range; it is merely necessary for sound at these extremes to be louder in order to be heard (Table 1.1).

While pure tones such as the one described above can easily be generated, most real sounds are not pure tones. Some, such as the notes produced by musical instruments, are a mixture of a relatively small number of frequencies which are related to each other (Figure 1.4). For example, a note which is based on 440 Hz may also contain components at 880, 1320, 1760 Hz and so on. Nonmusical

Table 1.1 Frequency and hearing

Frequency	Significance
20 Hz	Normally taken to be the lowest audible frequency
100 Hz	The mains hum emitted by a badly designed transformer or audio system
30–4000 Hz	Range of a piano keyboard
250–1000 Hz	Range of a female singing voice
125–6000 Hz	Range of frequencies present in speech (male voice)
200–8000 Hz	Range of frequencies present in speech (female voice)
20 000 Hz	Normally taken to be the highest audible frequency

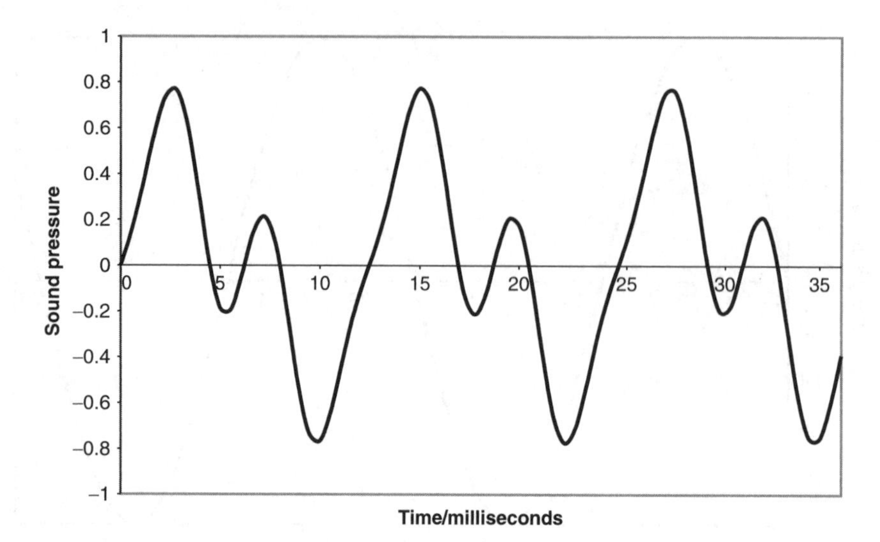

Figure 1.4 A musical note.

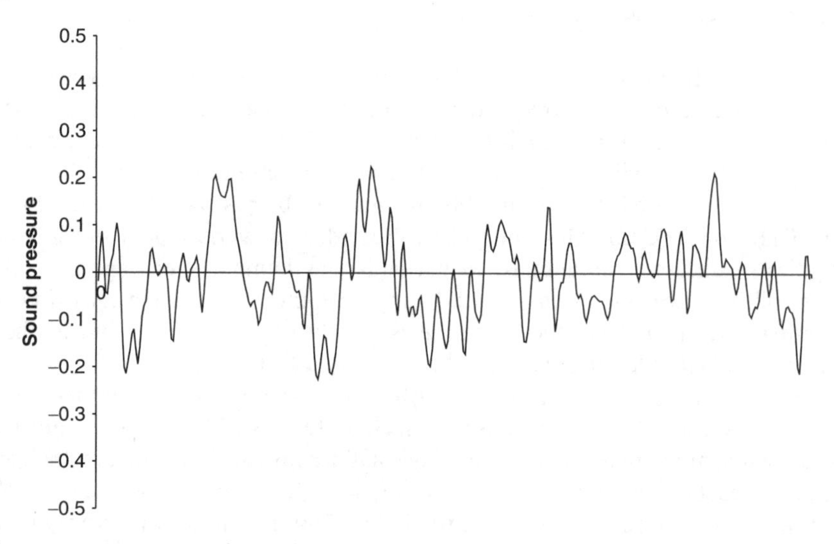

Figure 1.5 A nonmusical sound.

sounds, including most of those to which people are exposed in the course of their work, do not obviously possess the property of pitch (Figure 1.5). Investigation shows that these sounds consist of a mixture of a great number of frequencies which are not related to each other in any simple mathematical way.

As well as a frequency, a sound wave at a single frequency will also have a wavelength. The wavelength is the distance between successive peaks (or between successive troughs) of a wave, and like any other length it is measured in metres. Under normal conditions, audible sounds can have wavelengths varying from a few centimetres to a few metres. At 1 kHz, for example, the wavelength will be about 34 cm. At 100 Hz, the wavelength is around 3.4 m.

The wavelength and frequency of a particular wave are related by a simple equation:

$$v = f \times \lambda \tag{1.1}$$

where v is the velocity with which the wave travels (this varies slightly with temperature, but is about 340 ms^{-1} in air at room temperature); f is its frequency; and λ is its wavelength.

The above equation can be rewritten as

$$f = \frac{v}{\lambda} \tag{1.2}$$

in which case it is easy to calculate the frequency corresponding to a given wavelength. Alternatively, if it is required to work out the wavelength from a knowledge of the frequency, it can be expressed as

$$\lambda = \frac{v}{f} \tag{1.3}$$

Example

(a) What is the frequency of a sound wave with a wavelength of 2 m? (b) What is the wavelength of a sound wave with a frequency of 250 Hz? Assume the speed of sound is 340 ms^{-1}.

(a) $f = \frac{v}{\lambda} = \frac{340}{2} = 170\ \text{Hz}$

(b) $\lambda = \frac{v}{f} = \frac{340}{250} = 1.36\ \text{m}$

rms averaging

If we want to describe how loud a sound is, then we first have to produce from a waveform like the one described above, a number which describes the magnitude of the pressure fluctuations. Just averaging the deviations from normal pressure is of little use as the pressure spends as much time above the normal pressure as below it. The average would be zero. Instead, we normally get the measuring equipment to calculate the root-mean-square (rms) pressure.

This is always positive and varies, as you might expect, with the loudness of a sound; loud sounds produce higher rms values than quieter ones. Other sorts of averaging could have been used, but rms averaging is widely used by electrical engineers and is useful because it relates directly to the energy content of a sound wave.

The decibel scale

A sound consists essentially of a moving series of pressure fluctuations, and the normal unit of pressure is the pascal (abbreviated to Pa). However, it is not normal to measure sound in pascals; instead the decibel (abbreviated to dB) scale is used. The decibel scale is a logarithmic one, which compresses a large range of values to a much smaller range. For example, the range of sound pressures from 0.00002 to 2.0 Pa is represented on the decibel scale by the range 0 to 100 dB. Two justifications are normally given for using a decibel scale.

1. The range of values involved in measuring the amplitude of sound is inconveniently large.
2. The human ear does not respond linearly to different sound levels and the decibel scale relates sound measurement more closely to subjective impressions of loudness.

Neither of these explanations really stands up to scrutiny. We cope with larger ranges of values when measuring other quantities (length and money are just two examples of this). It is certainly true that our ears do not respond linearly to changes in sound pressure. In other words doubling the sound pressure does not double the apparent loudness of a sound. However, they do not respond linearly to the decibel scale either, so little has been gained in this respect by using a decibel scale. Whatever the original reasons for adopting a decibel scale, it is now used universally, so there is no alternative but to do so (Table 1.2).

The use of a logarithmic scale dates from the days before electronic calculators when many calculations were carried out with the help of a book of logarithms, or 'log' tables. As a result, logarithms were much more familiar to anyone who needed to carry out calculations regularly. Many fewer people are nowadays familiar with them. Fortunately, with the help of a calculator, decibel calculations can be carried out without any great understanding of how logarithms work. In other fields, different logarithms – called natural logarithms and abbreviated to either $\log_e$ or ln – are used. In workplace noise calculations, all logarithms will be the more familiar system based on the number 10. They are sometimes called 'logs to base 10', abbreviated to $\log_{10}$, log or simply lg.

Table 1.2 Everyday decibel levels

120 dB	Shot blasting enclosure
110 dB	Night club
100 dB	Operating position of wood planer
90 dB	Small engineering workshop
80 dB	Underground train
70 dB	Busy open-plan office
60 dB	
50 dB	Private office
40 dB	
30 dB	Rural location at night

The decibel scale for measuring sound levels is defined by the equation:

$$L_p = 20 \times \log_{10}\left(\frac{p}{p_0}\right) \quad (1.4)$$

where L_p is the sound pressure level; p is the rms sound pressure; and p_0 is a reference pressure which has the value of 2×10^{-5} Pa.

Example

What is the sound pressure level when the rms pressure fluctuation is 0.5 Pa?

$$L_p = 20 \times \log\left(\frac{p}{p_0}\right) = 20\log\left(\frac{0.5}{2 \times 10^{-5}}\right)$$
$$= 20 \times \log(25\,000)$$
$$= 20 \times 4.398$$
$$= 87.96 \approx 88.0\,\text{dB}$$

Sometimes, it is necessary to work out the rms sound pressure from a given sound pressure level. In this case, the subject of the above equation can be changed to give:

$$p = p_0 \times 10^{\frac{L_p}{20}} \quad (1.5)$$

Note that whereas sound pressure level is traditionally abbreviated to SPL, and this abbreviation is still commonly seen. This book follows modern practice in using the abbreviation L_p for sound pressure level.

Addition of noise sources

Because the decibel scale is a logarithmic one, it is not possible just to add together decibel quantities. Two noise sources, each of which individually results in a sound pressure level of 70 decibels (a typical level in a busy office) will not result in anything like 140 decibels (painfully loud) when operated together. In practice, the combined level is likely to be around 73 dB, and this is because a logarithmic method must be used to combine the decibel levels. This is sometimes called *decibel addition* – a phrase avoided here as it might be confused with ordinary addition.

Imagine that two machines are installed in a workshop. With machine number 1 switched on and machine number 2 switched off, the measured level is L_1. With machine 1 off and machine 2 on, the level is L_2. When both machines are switched on together, the combined level is given by:

$$L_p = 10 \times \log\left(10^{\frac{L_1}{10}} + 10^{\frac{L_2}{10}}\right) \qquad (1.6)$$

Here, L_p is used to denote the combined sound levels, while the levels due to each source on its own are denoted by L_1 and L_2.

With a calculator and a little practice, calculations such as this can be carried out reasonably easily.

Example

Two machines give individual sound pressure levels at a particular workstation of 86 and 88 dB, respectively. What will be the sound pressure level when they are both switched on together?

$$\begin{aligned} L_p &= 10 \times \log\left(10^{\frac{86}{10}} + 10^{\frac{88}{10}}\right) \\ &= 10 \times \log(3.98 \times 10^8 + 6.31 \times 10^8) \\ &= 10 \times \log(10.29 \times 10^8) \\ &= 90.12 \\ &\approx 90\,\text{dB} \end{aligned}$$

Modern calculators may give an answer in up to 10 digits. It is clearly pointless, and may be misleading, to copy all these into a report.

The convention is that the final significant figure quoted indicates the range of uncertainty of the value quoted. For example a calculator used for the above calculation offered the answer 90.12442603 dB. This has been rounded to 90 dB, implying that the 'true' value is believed to be somewhere between 89.5 and 90.5 dB. To state the answer as 90.1 dB would have implied that the 'true' answer has been narrowed down to between 90.05 and 90.15 dB.

It is not normally possible for the final answer to be more precise than the initial data. The question quoted the sound pressure level due to the individual machines to the nearest decibel, so the combined value cannot be known with any more precision than this. 90 dB is the correct form of the answer.

In sound measurements generally, and particularly in noise at work exposure assessments, it is not normally justified to claim greater accuracy than is implied by quoting final results to the nearest decibel. Sound level meters measure levels to the nearest tenth of a decibel and where calculations are involved significant errors can accumulate if this rounding is carried out too soon. It is best to record all readings as they appear (i.e. to a tenth of a decibel) and to carry this level of precision through any calculations that follow. The final result, though, should normally be rounded to the nearest decibel.

Equation 1.6 can also be amended to carry out other calculations. For example, if more than two sound pressure levels are to be combined it simply becomes

$$L_p = 10 \times \log\left(10^{\frac{L_1}{10}} + 10^{\frac{L_2}{10}} + 10^{\frac{L_3}{10}} + \ldots\right) \tag{1.7}$$

where L_3 is the sound pressure level when the third machine operates on its own. As many similar terms can be added as is necessary.

If several similar machines are in operation, then L_1 will be equal to L_2 etc., so that for n similar machines:

$$L_p = 10 \times \log\left(n \times 10^{\frac{L_1}{10}}\right) \tag{1.8}$$

Example

A machine is the main noise source in a workshop, and when it is operating the sound pressure level in the workshop is 84 dB. What sound pressure level would be expected if a second, identical machine were installed

$$L_p = 10 \times \log(2 \times 10^{\frac{84}{10}}) = 87\,\text{dB}$$

Finally, it is sometimes necessary to 'subtract' one level from another. For example, it may not be possible to run machine 2 without also running machine 1. In this case, the level due to source 2 alone can be calculated as follows:

$$L_2 = 10 \times \log\left(10^{\frac{L_p}{10}} - 10^{\frac{L_1}{10}}\right) \tag{1.9}$$

where L_p is once again the combined sound presure level and L_1 and L_2 are the sound pressure levels due to the individual noise sources.

It is sometimes quicker, and some people find it easier, to combine decibels by means of a graph such as the one in Figure 1.6. There are also 'look-up' tables which do the same job as the graph.

Suppose you have measured L_1 and L_2 as above. Subtract the lower of the sound pressure levels from the higher one, and find this difference on the

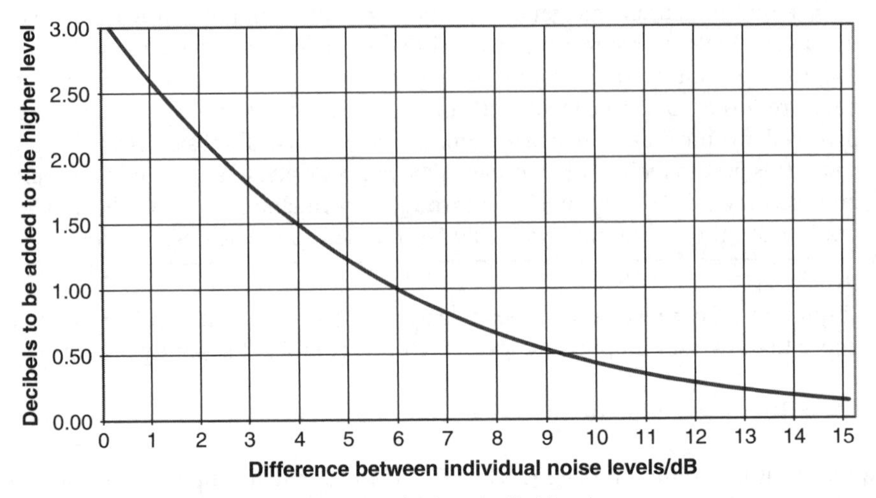

Figure 1.6 A graphical approach to combining decibel levels.

horizontal axis. Follow this value up to the curve and then read off the corresponding level on the vertical axis. This value must then be added to the higher of the individual sound pressure levels to give the combined sound pressure level.

> ***Example***
>
> Two machines give individual sound pressure levels at a particular workstation of 86 and 88 dB, respectively. What is the sound pressure level when they are both switched on together?
>
> $88 - 86\,\text{dB} = 2\,\text{dB}$.
>
> 2 dB on the horizontal axis corresponds to 2 dB on the vertical axis. So the combined SPL is $88 + 2 = 90\,\text{dB}$.

This is not quite as accurate as the mathematical method, but is adequate for most purposes. Subtraction is difficult this way and has to be done by trial and error. Multiplication can only be done by repeated addition.

There are a few decibel calculations which can be done without recourse to either of the methods described so far. Have a look at the graph for combining decibels.

- If there are two sources, each of which individually produce the same sound pressure level, then when both are switched on together the overall level will be 3 dB above either of the individual sources. This situation – or something approximating to it – is common enough in practice to be worth remembering.
- If there are 10 sources which individually produce the same sound pressure level, then the combined level will be 10 dB higher than the sound pressure level due to each source on its own.

- When two sound pressure levels are combined which differ by 10 dB or more, then the combined sound pressure level is essentially the same as the higher one on its own and the lower one can simply be ignored.
- The above rules can be combined. For example, four similar machines produce a level 6 dB (i.e. 3 + 3) above the level produced by one machine alone, while 20 similar sources result in an increase of 13dB (10 + 3).

Example

A machine produces a level at a particular workstation of 92dB. What will be the level at the same workstation if another identical machine is installed adjacent to the first?

Two identical machines will produce 92 + 3 = 95 dB

Wave properties of sound: reflection, absorption, refraction and diffraction

Different types of wave share a number of properties, and some of the properties of sound waves can be illustrated by comparing them to light waves.

Many surfaces will *reflect* sound waves; some will reflect virtually all the sound energy which strikes them, while others will *absorb* a significant portion of the energy which strikes them. The proportion of energy absorbed by a surface normally depends on the frequency of the sound. A reflection may be a *specular* one – as when light is reflected from a mirror – or it may be *diffused* by an irregular surface as when a matt white surface reflects a similar proportion of the light which strikes it. Smooth surfaces reflect sound waves in a similar way to light being reflected from a mirror (Figure 1.7). A wave striking a reflective surface at a particular angle will be reflected at a similar angle. A curved surface can focus sound energy in the same way as a concave mirror does with light.

Refraction occurs when a wave passes from one medium to another in which it travels at a different velocity. It accounts for the fact that water is normally deeper than it appears to an observer above the surface. Refraction can be important when investigating environmental noise, but it is of little importance in the workplace.

In the absence of obstructions, waves tend to travel in straight lines. *Diffraction* occurs when a wave encounters a sharp-edged obstruction. It accounts for the fact that sound can still be heard, even when the source is not directly visible. Diffraction properties depend critically upon the wavelength (light waves, by contrast, have very short wavelengths, so their diffraction is not often obvious). Thus low frequency sounds, which have long wavelengths, are more easily diffracted over and around obstructions than are higher frequency sounds (Figure 1.8).

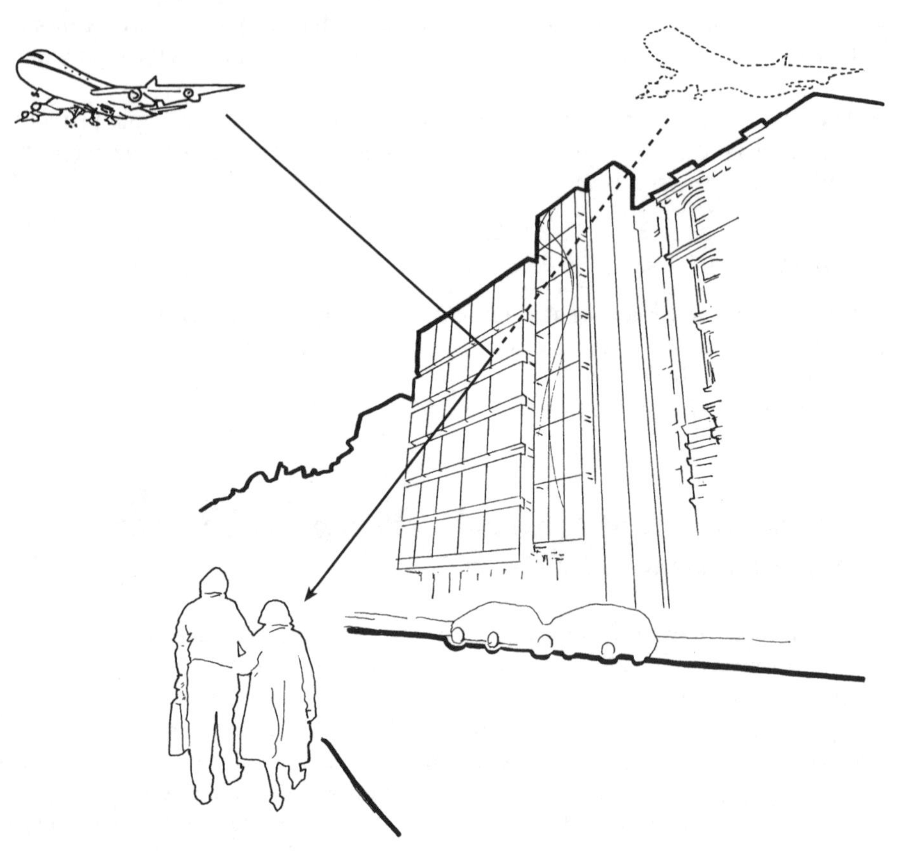

Figure 1.7 Reflection of sound waves. The aircraft appears to be in a different position due to the reflections from a large building.

Standing waves

Where there are large parallel reflecting surfaces, standing waves can be set up at particular frequencies, and these standing waves may lead to very high sound levels at particular spots within the room. This happens because sound energy is reflected backwards and forwards between two or more surfaces in such a way that at particular frequencies the incident and reflected waves are in phase with each other. This can lead to unexpectedly high sound pressure levels at particular points in the room – called antinodes – and there will be corresponding points – called nodes – where the waves tend to cancel each other out. The practical consequence is that noise measurements made at two points a few tens of centimetres apart may be very different, and as a result an inaccurate assessment may be made of an individual's noise exposure.

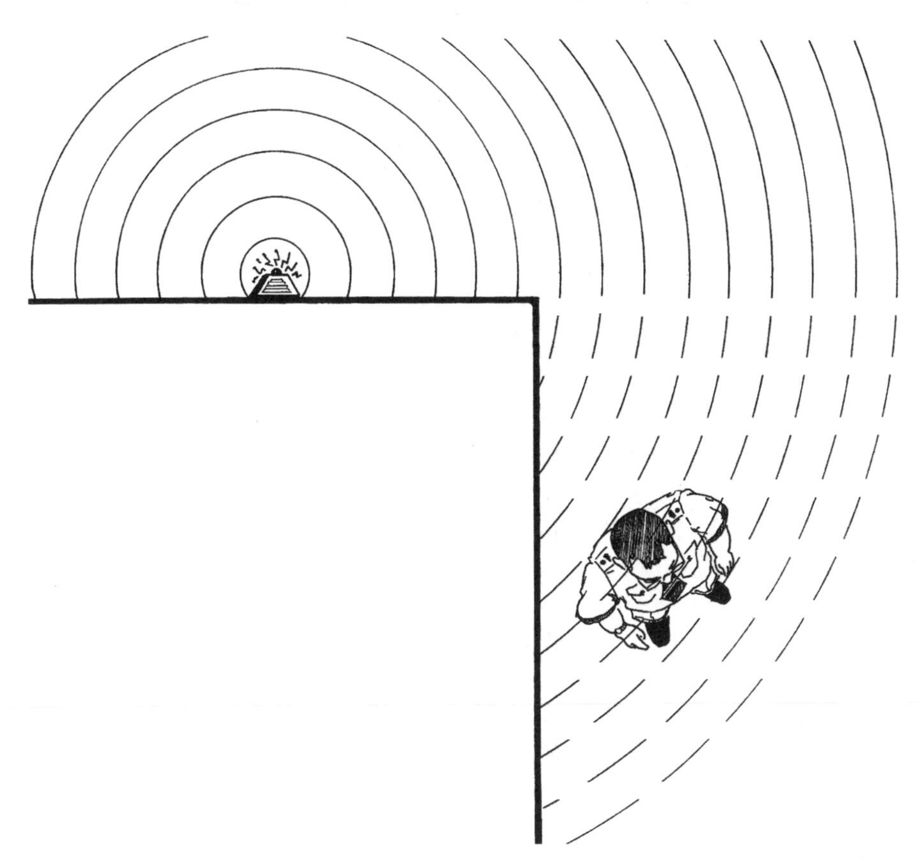

Figure 1.8 Diffraction of sound waves. The alarm is audible even though there is no direct path for the sound to travel.

Sound power and intensity

Sound pressure is measured at a particular point and may result from several sources of sound. There are two other quantities which can be of interest.

Sound power is the total amount of sound energy emitted per second by a particular noise source. It is therefore a property of that noise source and will not depend on the environment in which it is placed. It may, though, depend on operating conditions. For example, the noise output of a circular saw will depend on whether it is running freely or being used to cut material. The decibel counterpart of sound power is called sound power level (abbreviated to L_W, SWL or PWL) and is the most useful quantity to use when one noise source is compared with another. An EU directive requires suppliers of machinery to measure and supply to prospective users a value for the sound power level of

Table 1.3 Sound pressure, sound power and sound intensity compared

	Sound pressure	Sound power	Sound intensity
Unit	pascal	watt	watt per square metre
Unit abbreviation	Pa	W	Wm^{-2}
Decibel quantity	Sound pressure level	Sound power level	Sound intensity level
Symbol	L_p	L_w	L_I
Alternative symbols	SPL	SWL, PWL	SIL
Measurement: facilities required	Measured with a sound level meter	Normally measured in a special test chamber	Measured with specialist equipment
Measurement issues	Measured at a particular position	A property of one noise source	Measured at a particular position. Direction information required
Significance for noise at work	The basis of noise exposure assessments	Used to compare noise sources. Used to predict sound pressure levels	

their products. If they quoted a sound pressure level instead, then the measuring position and its surroundings would need to be described in detail for the information to be meaningful.

Sound intensity is the amount of sound power flowing across a particular imaginary surface with an area of $1 m^2$. It is measured in units of watts per square metre (Wm^{-2}). Its decibel counterpart is sound intensity level, and it is measured in some advanced acoustical investigations. The main reason for mentioning sound intensity here is that the phrase is often wrongly used when sound pressure level is meant. Because it has a very specific and different meaning it is important not to use the term loosely (Table 1.3).

Free field and reverberant sound

It was stated earlier that sound waves in a free field behave in an easily predictable way. Where reflecting surfaces are present, and in particular inside buildings with many surfaces which reflect sound well, the situation is much more complicated.

In a reverberant space, as many workshops are, the sound energy arriving at a particular point will be a combination of sound energy arriving directly from nearby machinery and the reverberant field present throughout the workshop. This has consequences for noise control within the workshop. It may not be enough to reduce noise from nearby machinery without also attempting to control the reverberant contribution.

Another important difference between free field and reverberant sound is that in a free field, the sound pressure level can be expected to fall continuously as the distance to the receiving point increases. For a small source, it is often found that in the open air, the resulting sound pressure level will fall by 6 dB every time the distance doubles. In a reverberant field such as can be expected indoors, the sound pressure level may level out once the receiver is a couple of metres away, and any further benefit from increasing the distance may be very small.

2

Human response to noise

The ear

When Vincent van Gogh cut off his ear, he probably lost some of his ability to locate the source of sounds in the vertical plane. The rest of his auditory faculties would not have been greatly affected, because the pinna, which to artists and others *is* the ear, is separated from the rest of the hearing organ. Because the hearing organ is so sensitive, it is also susceptible to damage as a result of physical contact with objects in the environment. It is located inside the skull at the end of a short tube called the auditory canal or meatus. Typically, in an adult, this is around 25–30 mm long and has a diameter varying between 5 and 8 mm.

The meatus is an essential part of the hearing apparatus. When it is blocked by wax or by an ear plug, the ear's sensitivity can be considerably reduced. When it is functioning normally, it will pass sound waves across the full audible range, but it tends to amplify to some extent those sounds with frequencies in the region of 3 kHz. The meatus ends at the tympanum, or ear drum. This is a membrane stretched across the inner end of the ear canal, and it tends to move in response to pressure waves arriving along it. It marks the boundary between the outer and middle ears (Figure 2.1).

The middle air is a cavity filled with air which contains three small bones known as the ossicles. Individually they have the latin names *malleus*, *incus* and *stapes*. These translate into English as hammer, anvil and stirrup. They join together to form a set of levers, with one end of the hammer in contact with the ear drum. As the ear drum moves in response to incoming sound waves, this movement is transmitted to the hammer. At the other end of the chain, the far end of the stirrup rests on the oval window which forms the boundary with the inner ear. Although the middle ear is normally isolated from the atmosphere, the eustachian tube joins it to the throat, and this is opened during swallowing to

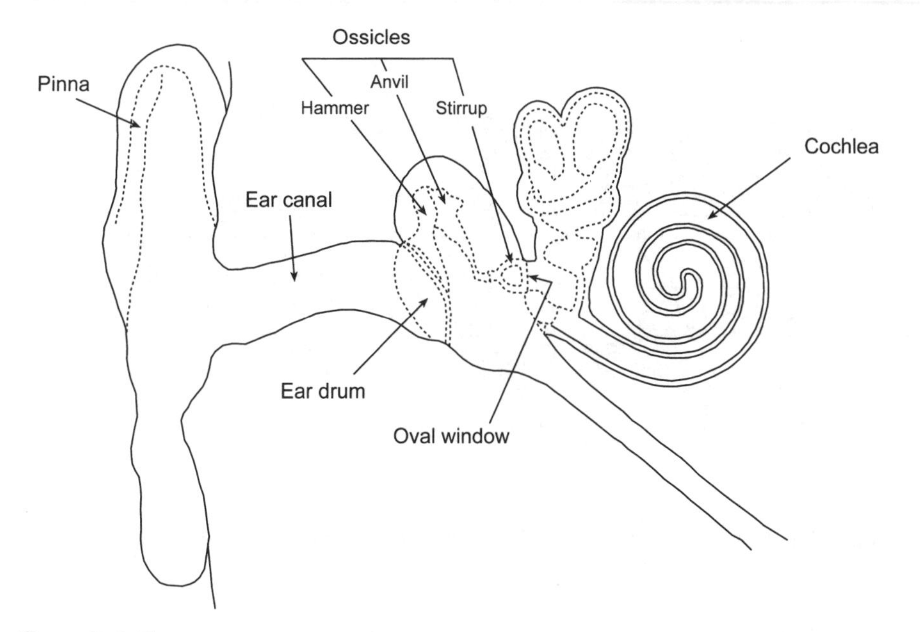

Figure 2.1 The ear.

allow the pressure inside the middle ear to become equal to the atmospheric pressure outside.

The inner ear is full of fluid, and the function of the ossicles is, by acting as a set of levers, to improve the efficiency with which the energy in the sound wave is transmitted to this fluid. This means that we can be sensitive to very quiet sounds, but when loud noises are present it can be a disadvantage for energy to be transferred too efficiently to the delicate inner ear. In this case muscles in the middle ear can act to reduce the efficiency of the process and thus, to some extent, protect the inner ear from damage.

Part of the inner ear detects movement and is important in maintaining balance. The section we are concerned with, though, is the cochlea. This is a double tube coiled up into a snail shape. The tubes are joined at the apex of the snail, and the base of one connects via the oval window with the stirrup bone, while the base of the other ends in the round window. They are separated by the basilar membrane, which is made to vibrate by the incoming sound waves. Different parts of the basilar membrane vibrate in response to different incoming frequencies, and this vibration is detected by hair cells within the membrane. These hair cells stimulate nerve endings to send electrical impulses to the parts of the brain which process auditory information (Figure 2.2).

The outer and middle ears are still to some extent exposed to external influences and can be damaged – for example by infections. They are also relatively simple mechanical systems which can be accessed by surgeons to correct defects.

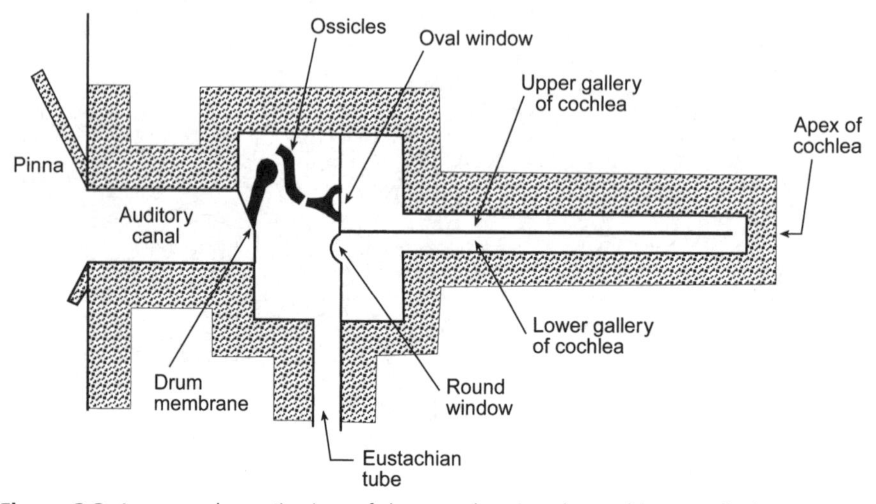

Figure 2.2 A more schematic view of the ear, showing the cochlea unrolled.

The inner ear is less exposed to damage (other than by the sound energy which it exists to detect) but there is little that can be done to correct any damage to it and the nervous system to which it is connected.

Hearing

The way in which human beings perceive and respond to sounds is a very involved subject indeed, and it is fortunately not necessary to deal here with most aspects of it. Some aspects of psychoacoustics, as it is called, relate to the way in which sound is measured and are therefore directly relevant to workplace noise assessment.

The range of frequencies audible to human beings is normally quoted as being from 20 Hz to 20 kHz (i.e. 20 000 Hz). However, this is a simplification for two reasons:

- There is a great deal of variation between individuals, and
- Our ears are not equally sensitive to all frequencies. It is necessary to make a great deal of noise at the ends of the frequency range for a sound to be audible. At 1 kHz the average young person can just detect a sound at zero decibels.

The sensitivity of human beings to sounds of different frequencies at different levels has been studied extensively, and as a result a series of equal loudness curves has been established. These answer the question 'How loud does a sound at each frequency have to be in order for an average individual to judge that it is as loud as a sound of x decibels at 1000 Hz?' When the figures that emerge as a set of answers to this question are plotted, they form one of a

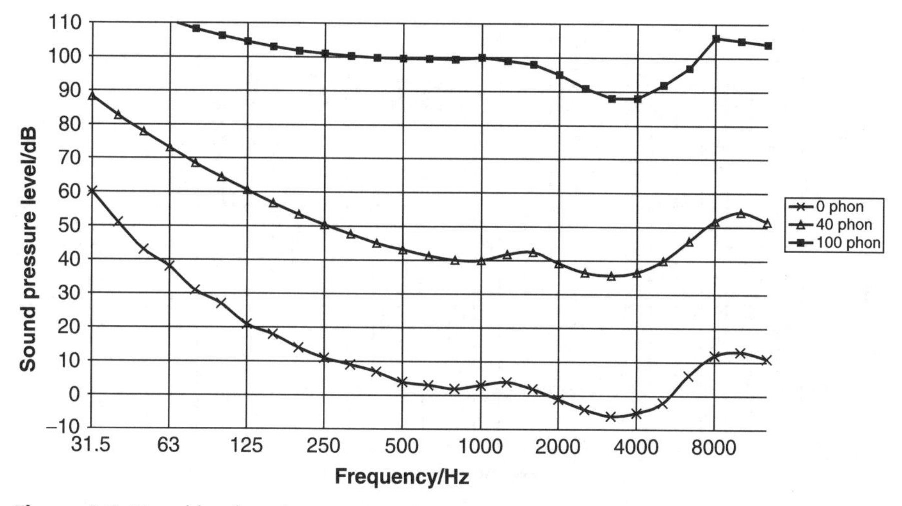

Figure 2.3 'Equal loudness' curves.

set of curves known as the equal loudness curves. Three of these are shown in Figure 2.3. The 'zero phon' contour is also known as the threshold of hearing, as at each frequency it is the level at which a sound is just audible to the average individual.

The phon is the unit used to assess subjective loudness in this way. The sounds at different frequencies which are judged to have the same loudness as a sound of 40 dB at 1 kHz all have a loudness of 40 phons. When these results are plotted for any one individual, they are unlikely to be as smooth as the curves shown, which result from averaging a number of individuals.

The measurements described above are subjective ones; that is they rely on human beings making a judgement and responding honestly to the researcher. Nevertheless, measurements on small groups of subjects seem to give consistent results. Most measurements on human hearing, including the hearing tests described below, depend in a similar way on the reliability of subjective responses.

When measuring noise levels, we normally want to arrive at a figure that corresponds very closely to the apparent loudness of a sound as heard by a human being. In order to do this the measuring equipment must take into account the varying response of the ear to different frequencies. This can be done by building into the measuring instrument a circuit which 'weights' the different frequencies in such a way that a subjectively loud sound will result in a high decibel reading and a quieter sound will result in a lower reading.

The equal loudness curves above show not only that the human response to sounds depends on frequency, they also show that the frequency dependence changes with level. Specifically, at 100 dB the ear responds more uniformly to different frequencies than is the case around the threshold of hearing. In the middle of the twentieth century, when these matters were first studied in detail, it

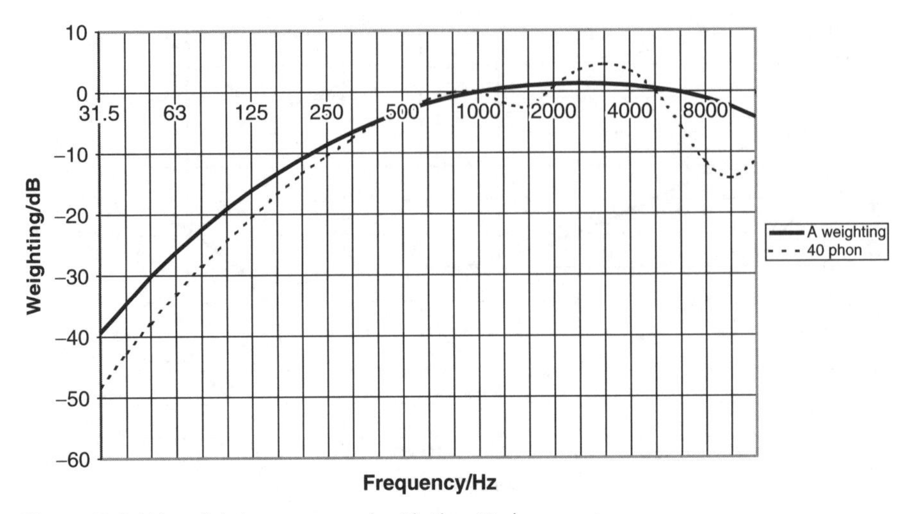

Figure 2.4 'A' weighting compared with the 40 phon contour.

was thought that it would be necessary to use different weighting systems depending on how loud the sound was. Three frequency weighting systems were defined for use when measuring sounds in different level ranges, and were named the A, B and C weightings (later on a D weighting was added). It proved to be a cumbersome system and the B, C and D weightings fell into disuse as it became clear that the A weighting was adequate for many purposes. The C weighting has re-emerged and is used for situations where it is required to attach more weight to the low frequency components of noise. The most important such use is for peak noise levels in workplace noise assessment.

The A weighting curve as shown in Figure 2.4 is based on the 40 phon contour. It is upside down compared to the equal loudness curves, since we are now answering a different question: 'How important is the contribution of each frequency in deciding the loudness of a particular sound?' instead of 'how much noise do we have to make at a particular frequency for it to sound as loud as a 1 kHz tone at a particular level?' It is also rather simpler than the equal loudness curve since it was defined in a time when the electronic circuitry required for a more complicated curve would have been expensive and bulky.

Audiometry

Routine testing of the hearing ability of employees in a noisy workplace, or indeed of patients referred to a hospital department, requires equipment and procedures which are easily achievable but which yield consistent, repeatable results. Audiometry – the process of testing hearing ability – and audiometers – the instruments which are used to carry out the test – concentrate on testing the subject's hearing threshold. That is to say at each of the test frequencies they determine the level of the quietest sound which the subject can hear.

Various international standards are involved in the audiometry process:

- Part 1 of IEC 60645 describes the characteristics of audiometers to be used for various types of hearing test;
- ISO 8253 concerns itself with the test environment and the procedures to be used by the operator;
- Part 1 of ISO 389 defines the levels to be assumed to be normal for a young person with no hearing impairment.

The subject is seated in a quiet environment – either in a sound-proofed booth or in a naturally quiet room – and is fitted with headphones which are adjusted in size so that they are correctly positioned over the ears. The operator explains how the test will proceed, and how the subject is expected to respond to the sounds that will be presented through the headphones. It is important that the subject cannot see the operator or the audiometer during the test.

Both manual and automatic recording audiometers are in use. The operator of a manual audiometer will select the frequency and level of the sound and will press a button to make a sound in the appropriate headphone. The subject is instructed to press their response button whenever a tone is heard. The operator will normally test one ear at a time, and will work through the frequencies and levels in a methodical way, recording for each frequency the lowest level to which the subject consistently responded.

An automatic recording audiometer presents the frequencies in turn to each ear. The subject is told to press the response button and to keep it pressed for as long as the pulsed tone is audible. While the response button is pressed, the audiometer will gradually reduce the level until it is no longer audible and the subject releases the button. Once the button is released, the audiometer will gradually increase the level until it is once more audible and the subject presses the button once again. This process is repeated a few times, after which the levels at which the button was pressed and released are averaged to determine the hearing threshold of that ear at that frequency. The operator monitors the automatic audiometer and watches for any anomalies in the responses; it may be possible to alter the sequence to check on unexpected results.

Tests using an automatic recording audiometer are quicker, and results depend less on the operator's technique. Manual audiometers are cheaper, and the operator has more control over the test. This may be important if the subject has difficulty understanding what is required, or if an attempt to falsify the results is suspected.

The audiometer measures hearing threshold levels relative to what is agreed to be the average hearing level of a group of young people with no known hearing problems. Thus an individual whose hearing exactly matched this norm would record a hearing level of 0 dB at each frequency. However, there is a great deal of variation even in the hearing ability of those young people who are thought to have 'normal' hearing, and levels in the range $\pm$ 20 dB would be regarded as

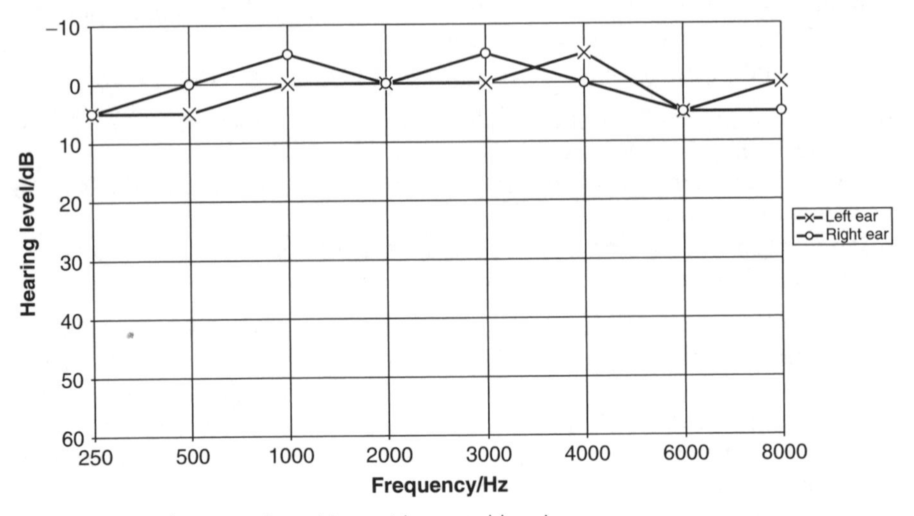

Figure 2.5 Audiogram of a subject with normal hearing.

within the normal range. If at any frequency the subject is more sensitive than the norm, then their threshold at that frequency will be negative. If they are less sensitive than the norm, their hearing threshold is positive. The worse the hearing, the higher the threshold. It is thought, however, that audiograms are easier to read if good hearing is represented by points at the top of the graph. The vertical axis is therefore upside down in the sense that the lowest numbers appear at the top of the axis, as shown in Figure 2.5.

As individuals age, their sensitivity to high frequency sounds is progressively reduced (Figure 2.7). As well as starting from slightly different levels as young people, different individuals' hearing will deteriorate at different rates. Nevertheless, it is possible to plot average hearing levels for individuals at different ages (Figure 2.6). Age-induced hearing loss is called presbycusis or presbyacusis.

Prolonged, repeated exposure to high noise levels will also cause a loss of hearing sensitivity. In this case, it is normally most obvious at 4 kHz (although sometimes there is a greater loss at 6 kHz). If a subject's hearing is better at 8 kHz than it is at 4 kHz, then there will normally be a history of noise exposure (Figure 2.8).

Leisure noise exposure can sometimes cause significant noise loss in individuals who take part in shooting or who attend (or take part in) musical performances regularly. However, it much more commonly results from work activities.

Both presbycusis and noise-induced hearing loss are forms of sensorineural hearing loss. Sensorineural loss is caused by damage to the cochlea and/or the auditory nerve, and for practical purposes cannot be reversed by surgical intervention. Conductive hearing loss occurs when the damage is in the middle and/or outer ears, and there are a number of surgical techniques which can reduce or reverse this type of loss. It is not always obvious from the audiogram which type

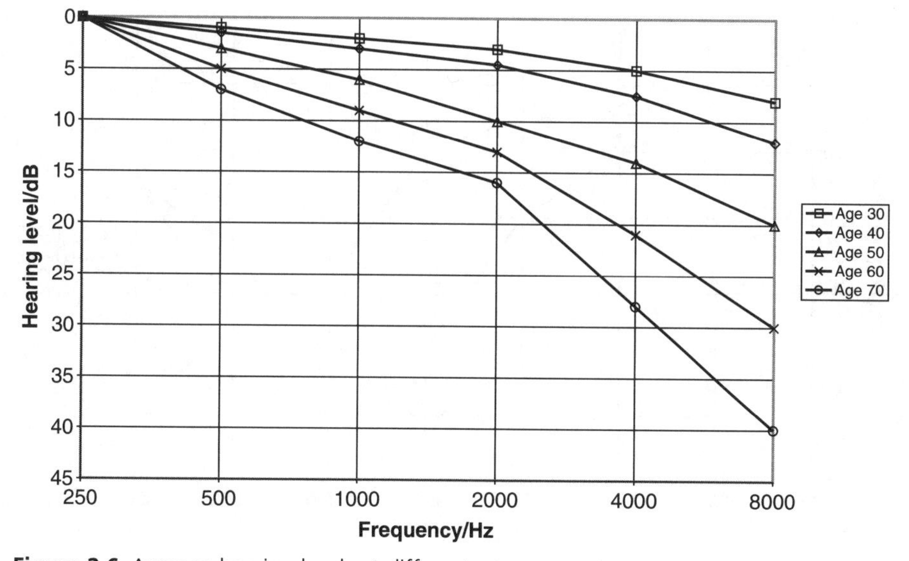

Figure 2.6 Average hearing levels at different ages.

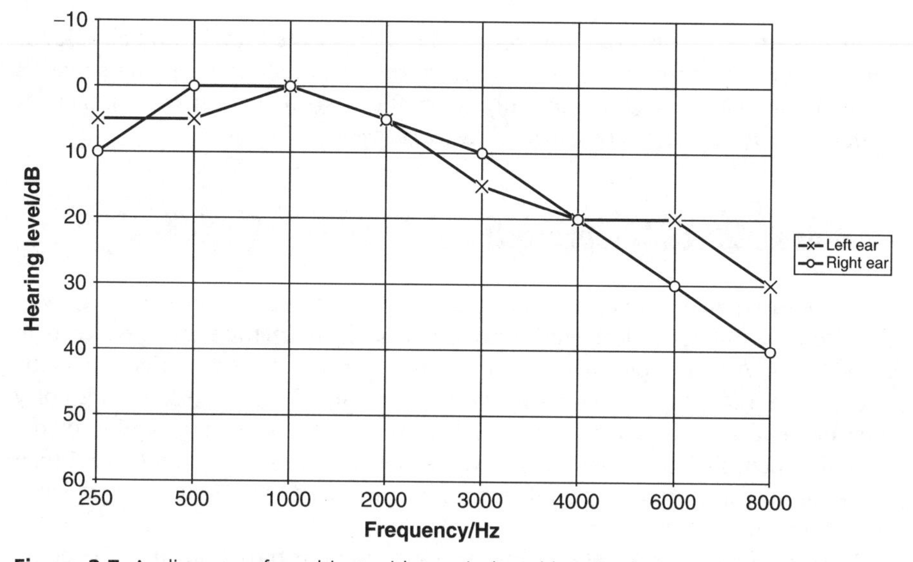

Figure 2.7 Audiogram of a subject with age-induced hearing loss.

of loss has occurred, but they can be distinguished by using an audiometer fitted with a bone conductor instead of a headphone. This transfers energy directly to the mastoid bone, from which it can reach the cochlea without passing through the outer and middle ears. A normal bone conduction audiogram would suggest that there is no damage to the inner ear, even if the air conduction audiogram is

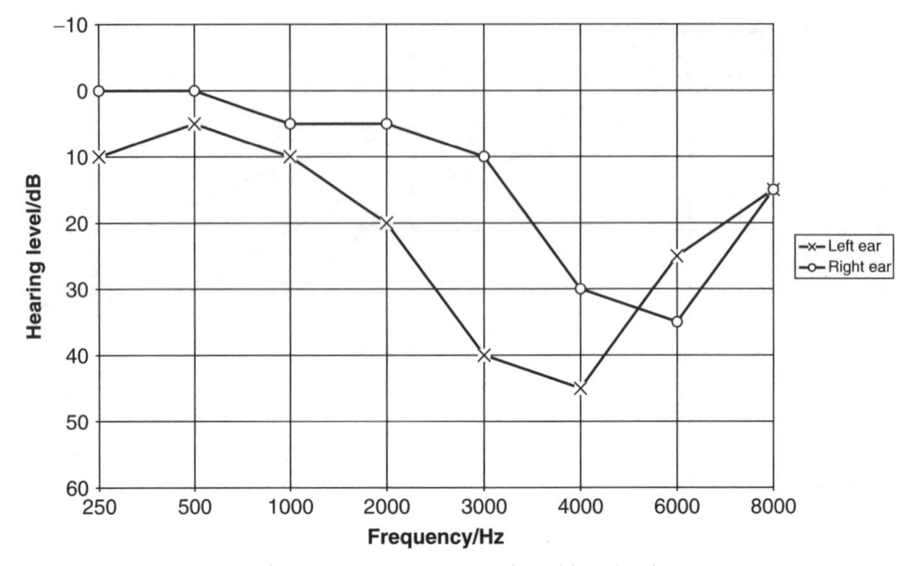

Figure 2.8 Audiogram of a subject with noise-induced hearing loss.

showing a degree of hearing loss. On a less sophisticated level, a quick test for sensorineural hearing loss is to place a vibrating tuning fork on the subject's skull. If this can be heard normally, then the first indication is that there is conductive rather than sensorineural hearing loss.

Noise-induced hearing loss

Many of those who have been exposed to noise for several years will also show at least some age-induced loss, and it can be difficult to separate the two effects in some cases. An example of this – showing the hearing pattern that might be expected in a 60-year-old individual who had spent 40 years working in a noisy environment – is shown in Figure 2.9. If the existence of noise-induced loss is to be identified in either an individual or in a group of employees, then allowance must be made for that portion of the measured hearing loss which is due to aging.

The UK Health and Safety Executive (1995) recommends a scheme – based on the international standard ISO 1999 – for categorizing hearing test results, based on the threshold levels and the age of the subject. First the hearing levels at three low frequencies (0.5, 1 and 2 kHz) are added together. Separately, the hearing levels in three high frequency bands (3, 4 and 6 kHz) are added. The two totals are compared with a threshold level – dependent on the subject's age – above which it is recommended that the subject is warned of an apparent loss of hearing, and a higher threshold – also age-dependent – above which referral to a medical

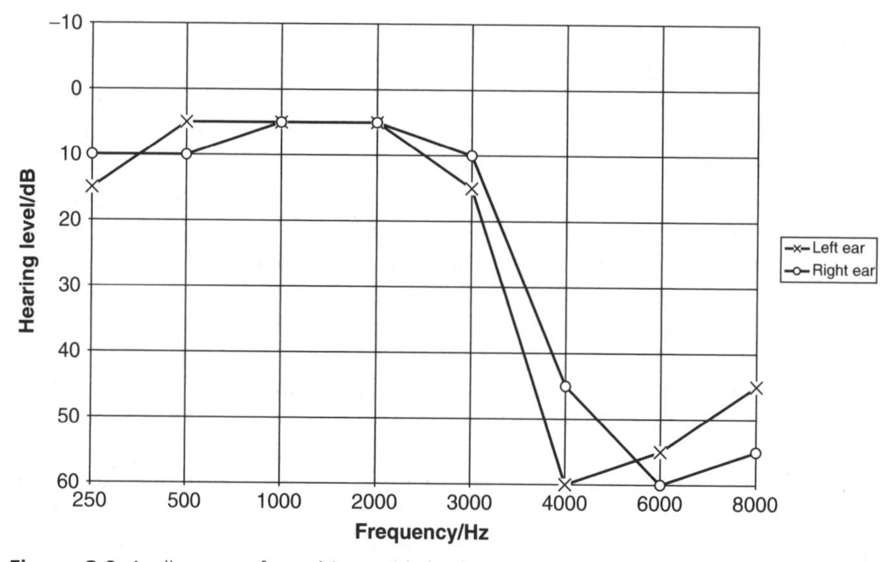

Figure 2.9 Audiogram of a subject with both age-induced and noise-induced hearing loss.

specialist should be made. Other findings which should result in referral to a specialist are:

- A change of more than 30 dB in either total since the previous test
- A difference between the two ears of more than 45 dB (for the low frequency total) or 60 dB (for the high frequency total).

When an employee has suffered a hearing loss as a result of occupational noise exposure, it has to be remembered that they will subsequently suffer a high frequency loss as they age, which will be added to the noise-induced loss. In normal conversation there is a great deal of redundant information which means that even if some sounds or words are not heard, the human brain will normally be able to reconstruct the entire conversation from the information it has. However, where communication conditions are poor (in conditions of high background noise, say, or on the telephone) it can be very much more difficult for a person with a moderate amount of hearing loss to communicate effectively.

Temporary threshold shift and tinnitus

When the ear is exposed to high noise levels, the stapedius muscle in the middle ear acts to reduce the transfer of sound energy to the inner ear. This is a reflex which operates within a few seconds, and recovers only very slowly. The result is that those who have recently been exposed to high noise levels, even for a short period, will suffer from a degree of temporary threshold shift. This is often very noticeable to those who have been listening to loud music. It is also known to disc jockeys, who gradually increase the music's volume as the evening progresses in order to offset the apparent volume reduction as the audience's sensitivity to noise is reduced.

If a hearing test is carried out while a temporary shift is present, the results are very similar to those who have permanent threshold shift. The difference can only be established by retesting after a period of 2–3 days during which no significant noise exposure has occurred. Traditionally, industrial nurses would carry out audiometric testing on Monday mornings after the ears had had the weekend to recover from any temporary shift. Nowadays, employees whose noise exposure at work is controlled may willingly expose themselves to high levels of noise in their leisure activities, and simply carrying out the tests on a Monday morning is not a reliable way of ensuring the results are unaffected by temporary threshold shift. It is important when testing hearing to use a questionnaire to establish whether the results are likely to be affected by recent noise exposure.

Tinnitus is a condition in which the sufferer can hear a variety of noises which are not caused by external noise. It frequently accompanies temporary threshold shift, when it is also a temporary phenomenon. As a result of long-term exposure it can become permanent. It can also have a number of causes not related to noise exposure.

The nature of the noises heard varies between individuals, both in terms of the nature of the noise heard and also how often it is present. Some sufferers will hear a rushing in their ears only when they are in a particularly quiet environment. Others may hear this kind of sound virtually continuously, or may be subject to more annoying noises such as clicks, pops and rattles. Sufferers frequently find it more difficult to come to terms with tinnitus than with hearing loss, which is less apparent to the sufferer.

The relationship between noise exposure and hearing loss

If hearing damage is to be prevented by limiting occupational noise exposure, then it is necessary to have some quantitative understanding of the relationships between sound pressure level, frequency, exposure time and the degree of damage caused. Having established that there is a risk to hearing, though, it would be unethical to refrain from taking all reasonable measures to prevent it. During the 1960s a great deal of work was done in the UK and the rest of the world to establish the relationships between noise exposure and noise-induced hearing loss. At that time it was relatively easy to find populations who had worked at one job, and been exposed to steady noise levels, for a number of years.

Since then, social mobility, changing patterns of employment, and indeed government action to limit noise exposure, have made it much harder to find large groups of workers whose noise exposure can be logged over several years.

Information on the precise relationship between the various factors influencing hearing damage is therefore incomplete, and a full understanding of the subject will never be achieved.

Full understanding is not required, though. What is needed is sufficient information to frame legislation and advisory procedures which are capable of being put into practice in such a way that occupational hearing damage is reduced and eventually eliminated, without also making essential industrial processes impossible or uneconomical to carry out. This is itself quite a demanding objective, and efforts to achieve it are further discussed in Chapter 4.

In studying the relationship between noise exposure and hearing loss a range of questions can be asked. It can be assumed that louder sounds will result in more damage than quieter ones, but more detailed questions include:

- Is there a sound pressure level below which there is no contribution to hearing damage?
- If so, then what is this level?
- If all noise contributes to damage then what is the trade-off between level and damage?
- Does an extended period of noise exposure do the same amount of damage as a series of shorter exposures at the same level?
- Are particular frequencies or ranges of frequencies significantly more damaging than others?
- Is there a link between the frequencies to which the ear is exposed and the frequencies at which hearing loss occurs?

The answers to these questions and other questions will all have consequences for the way in which noise exposure must be measured.

In the European Union, an approach to the assessment of noise exposure has emerged which uses the best available answers to these questions. Each of the assumptions listed below can be challenged, and together they represent a gross simplification of a very complicated area of knowledge. For the time being, they seem to offer a practical way forward to those working to reduce occupational hearing loss, and as stated above, that is the most that can be asked for.

1. All sound energy received by the ear will, in some degree, contribute to hearing damage.
2. The degree of damage is proportional to the amount of sound energy deposited in the ear. That means that a doubling of exposure time is equivalent to a 3 dB increase in sound pressure level. It also means that the total exposure time at a given level is important; breaking the overall time up into shorter periods has no effect.
3. The A weighting system correctly evaluates the contribution of different frequencies to hearing loss.
4. Very high sound pressures can cause damage which may not be reflected in an equal-energy assessment as described above. An additional limit on peak sound exposure can be used to prevent this.

In the United States, rather different conclusions have been reached, and as a result a rather different trade-off between sound pressure level and exposure time is used. This is based on the assumption that a 5 dB increase in level (rather than 3 dB) is equivalent to a doubling of exposure time. To add to the confusion, for some purposes in the United States 4 dB (rather than 3 or 5 dB) is assumed to be equivalent to a doubling of exposure time. Those carrying out noise exposure assessments in Europe need to be aware of these different practices in order to avoid being misled by procedures or instrumentation intended for American use.

The current European approach to the prevention of occupational hearing damage is based on the principles listed above. The issue is the subject of continuing debate as research into hearing damage continues. Given the difficulties of generating further large sets of data which can be used to refine our knowledge, it seems likely that for the foreseeable future this approach will continue.

Measuring noise

Time constants

Although some noise sources seem to be emitting steady noise levels, in practice it is normally found that the level is fluctuating to some extent. Some noise sources vary in level very obviously and by a great deal in a short time. The changes in level may contain important information, but they make it very difficult to measure because of the rapidly changing meter display. Two standard approaches were developed to allow readings to be made in a way that would give consistent results when used in different noise environments. One used a slow time constant of 1 s. This means that level fluctuations over periods of around a second would be smoothed out and an essentially steady noise source would generate a steady reading on a sound level meter. The fast time constant of 1/8 s, on the other hand, would allow a sound level meter to respond much more quickly to level changes although it would never be useful if a steady reading were required. Slow time constant was the more useful of the two when investigating workplace noise.

The impulse time constant was developed later as a way of estimating the effects of impulsive noises – hammering, explosions, etc. The characteristics of this type of noise make it much more intrusive to human beings than the sound level meter readings (either using fast or slow time constants) would suggest. However, the impulse time constant does not follow the equal energy principle used to assess the effects of noise exposure on hearing, and it is never used in workplace noise assessments. It is sometimes still used for some other purposes.

The peak time constant is rather different from the fast and slow time constants. It is discussed later.

When noise levels are fairly steady, it is possible to measure sound pressure level using the slow time constant and to arrive at an accurate noise exposure prediction by assuming that this level is maintained for a long period. But most sound level meters (other than the very cheapest models) are these days able to measure a true long-term average, called L_{eq}, and this is the preferred approach when noise levels tend to fluctuate, as they almost always do in real work situations.

The equivalent continuous level, L_{eq}

Like many other quantities, sound pressure levels could be averaged in a number of different ways, but the method normally used is to average, over a predetermined time period, the energy content of the sound. The result, L_{eq}, is the equivalent continuous level; the steady level which, if maintained for the whole of the time period, would have contained an equal amount of energy to that which was measured in the actual, fluctuating noise level. Clearly, its use in workplace noise assessments is consistent with the equal energy principle as described in Chapter 2. It also finds use in a number of other fields of noise assessment, including many assessments of the effects of environmental noise.

When A weighted noise exposure is of interest, then it is customary to make this clear by including an A in the subscript – hence L_{Aeq}. It is also possible to omit the 'A' and make it clear in another way that the measurement is an A weighted one 'an A weighted L_{eq} of 86.2 dB was measured . . . '. Because of its many applications L_{eq} measurements are made using different frequency weightings, so in one way or another the frequency weighting should be made explicit. Formerly, an A weighted sound pressure level measurement was indicated by adding (A) after dB – as in 'a sound pressure level of 86.2 dB(A)'. This obsolete form dates back to the days when A, B and C weightings were in common use, but is still used sometimes today.

Measurements of L_{eq} involve continuous averaging algorithms which are best left to microprocessor programmers. There are some simple situations, though, in which it is possible and desirable to calculate an L_{eq} with the aid of a scientific calculator. The simplest scenario is when a steady noise level is known to exist for a specified period, after which it abruptly changes to another specified level for another known time period, as shown graphically in Figure 3.1.

The L_{eq} for the entire period is calculated as follows:

$$L_{eq} = 10 \times \log \frac{1}{T}\left(t_1 \times 10^{\frac{L_1}{10}} + t_2 \times 10^{\frac{L_2}{10}}\right) \qquad (3.1)$$

where L_1 and L_2 are the sound pressure levels during the two subperiods, t_1 and t_2 are the time periods for which these levels are maintained, and T is the overall time period. T must equal the sum of t_1 and t_2, and moreover t_1, t_2, and T must all be measured in the same units, whether it is hours, minutes or seconds.

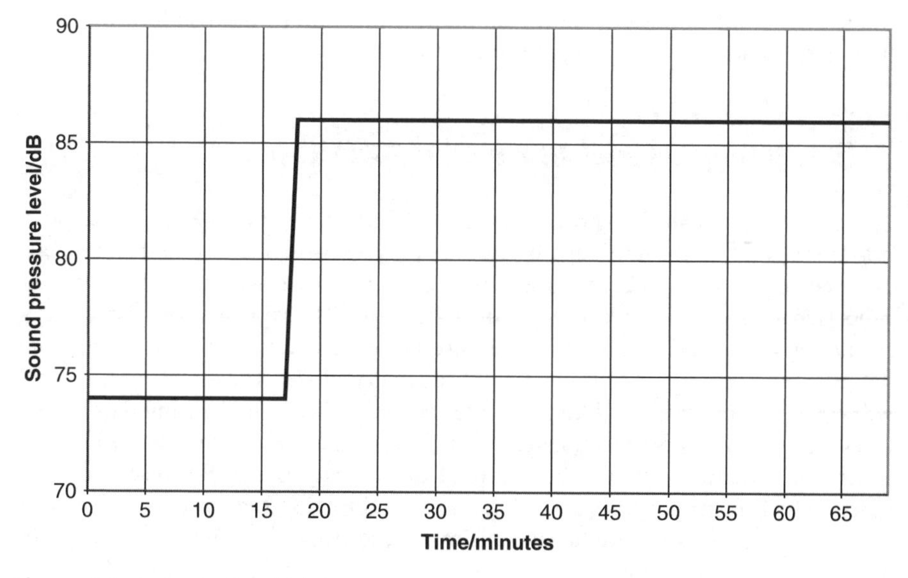

Figure 3.1 A simple time history.

Example

At a particular position the L_p is 86 dB for 3 h, and then 91 dB for 2 h. What is the L_{eq} for the entire 5 h period?

$$L_{eq} = 10 \times \log \frac{1}{T}\left(t_1 \times 10^{\frac{L_1}{10}} + t_2 \times 10^{\frac{L_2}{10}}\right) = 10 \times \log \frac{1}{5}\left(3 \times 10^{\frac{86}{10}} + 2 \times 10^{\frac{91}{10}}\right)$$
$$= 88.7\,\text{dB}$$

This equation for calculating the L_{eq} for a complete period can easily be amended to cover a wider range of situations. The simplest is when the two subperiods no longer represent periods for which the sound pressure level was unchanging. Instead, they can each be an L_{eq} measured by a sound level meter sampling a more complicated time variation for each of the two subperiods. All that is required is to put these measured L_{eq}s in the same equation where previously the steady sound pressure levels appeared.

$$L_{eq} = 10 \times \log \frac{1}{T}\left(t_1 \times 10^{\frac{L_{eq1}}{10}} + t_2 \times 10^{\frac{L_{eq2}}{10}}\right) \quad (3.2)$$

The next situation to which this equation can be extended is when there are more than two subperiods. As before, the data on each can be a steady SPL or, more commonly, it can be an L_{eq}.

$$L_{eq} = 10 \times \log \frac{1}{T}\left(t_1 \times 10^{\frac{L_{eq1}}{10}} + t_2 \times 10^{\frac{L_{eq2}}{10}} + t_3 \times 10^{\frac{L_{eq3}}{10}} + t_4 \times 10^{\frac{L_{eq4}}{10}} \ldots\right) \quad (3.3)$$

There is no limit to the number of subperiods that can be used so long as all the time periods add up to the overall time, T.

The daily personal noise exposure, $L_{EP,d}$

L_{eq} is useful for many purposes, but it is not adequate for assessing workplace noise exposure. Take the case of two employees who work together and are exposed to the same, steady, noise level. One works for 4 h every day, and the other frequently does overtime and clocks up 12 h exposure on those days. Assuming the sound level is the same throughout the day, this second employee has been exposed to three times the noise energy of the part-timer and, if these work patterns continue, is likely to suffer considerably more hearing damage.

To get round this problem, assessment of workplace noise exposure is done by means of a quantity called the personal daily dose, $L_{EP,d}$. Because it is used specifically for this purpose, and because A weighting is used for workplace noise assessment, $L_{EP,d}$ is in practice always measured using A weighting, and can be written $L_{AEP,d}$. The equation for calculating it from information about the exposure during subperiods looks very like the equation for calculating L_{eq}:

$$L_{EP,d} = 10 \times \log\frac{1}{8}\left(t_1 \times 10^{\frac{L_1}{10}} + t_2 \times 10^{\frac{L_2}{10}} + \ldots\right) \tag{3.4}$$

Here, the actual time period T has been replaced by a standard working day of 8 h. This standard working day is always 8 h, irrespective of the actual shift length. The only time it will change is if for some reason the time periods are all measured in minutes (or seconds) rather than hours. As with L_{eq} calculations, the time periods must all be expressed in the same units so in this case the 8 would be replaced by 480 (min) or 28 800 (s). As with L_{eq} calculations, the subperiod levels can be steady sound pressure levels, or they can be measured L_{eq}s. There can be as many subperiods as are required, and the individual time periods will normally add up to the total shift time, even though a standard 8-h day is used elsewhere in the equation.

In future, and in particular in the Physical Agents (Noise) Directive, $L_{EP,d}$ will be known as $L_{EX,8\,\text{hours}}$.

> ***Example***
>
> What is the daily personal dose, $L_{EP,d}$ for a worker exposed to the noise levels in the previous example?
>
> $$L_{EP,d} = 10 \times \log\frac{1}{8}\left(t_1 \times 10^{\frac{L_1}{10}} + t_2 \times 10^{\frac{L_2}{10}}\right) = 10 \times \log\frac{1}{8}\left(3 \times 10^{\frac{86}{10}} + 2 \times 10^{\frac{91}{10}}\right)$$
> $$= 86.7\,\text{dB}$$

Because the ways of working out L_{eq} and $L_{EP,d}$ look very similar, there ought to be a simpler way of converting one to the other, and so there is.

$$L_{EP,d} = L_{eq} + 10 \times \log\left(\frac{T}{8}\right) \tag{3.5}$$

and conversely,

$$L_{EP} = L_{ep,d} - 10 \times \log\left(\frac{T}{8}\right) \tag{3.6}$$

where T is the actual exposure time in hours.

Example

A group of employees is exposed to a steady sound pressure level of 91 dB. What will be the $L_{EP,d}$ of an employee who works for 5 h?

$$L_{EP,d} = L_{eq} + 10 \times \log\left(\frac{T}{8}\right) = 91 + 10 \times \log\left(\frac{5}{8}\right) = 89\,\text{dB}$$

In this example, the actual hours of work are less than 8, and so $L_{EP,d}$ is lower than L_{eq}. If the number of hours worked is greater than 8, then $L_{EP,d}$ will be greater than L_{eq}.

Describing short, noisy events

Sometimes a significant contribution to an employee's daily noise dose can be made by one or more short but very noisy events. Examples of this might include blasting in a quarry which takes place only once or twice a day, or aircraft noise affecting airport construction workers. A useful way of assessing noise doses in this case is to measure the noise contribution from a single event, using a quantity which is officially called the sound exposure level, and is abbreviated to L_{AE}. It also goes under a wide variety of other names and abbreviations:

- SEL, standing for
  - sound exposure level
  - single event level, or
  - sound equivalent level
- L_{eA}
- L_{AX}.

L_{AE} can be measured directly by many sound level meters and is the level which, if it were maintained for 1 s, would contain the same amount of sound energy as the actual event.

Once L_{AE} is known, the $L_{EP,d}$ for the day can be calculated:

$$\begin{aligned} L_{EP,d} &= L_{AE} - 10 \times \log 28\,800 + 10 \times \log n \\ &= L_{AE} + 10 \times \log n - 44.6\,\text{dB} \end{aligned} \tag{3.7}$$

where n is the number of similar events to which the employee is exposed, and 28 800 is the number of seconds in a standard 8-h working day, and $10 \times \log 28\,800 = 44.6$.

Many sound level meters allow L_{AE} to be measured directly. If one that does this is not available, then it can be calculated from the L_{eq} measured during one noisy event:

$$L_{AE} = L_{eq} + 10 \times \log t \tag{3.8}$$

where t is the time over which the noisy event was measured.

Example

A normally quiet machine automatically goes into a clean-down cycle four times a day. During one such cycle, an L_{eq} of 99 dB was measured at the operating position over a period of 1 min. Calculate the operator's $L_{EP,d}$.

First, calculate the L_{AE} for this event:

$$L_{AE} = L_{eq} + 10 \times \log t = 99 + 10 \times \log 60 = 117\,\text{dB}$$

Then use the calculated value of L_{AE} and the assumption that all clean-down cycles are the same to calculate the daily dose:

$$L_{EP,d} = L_{AE} + 10 \times \log n - 44.6\,\text{dB} = 117 + 10 \times \log 4 - 44.6 = 78.4\,\text{dB}$$

Peak and maximum levels

Various ways have been developed over the years of assessing the highest sound pressure encountered during a measurement, and several of them may be available on a sound level meter. Only one is important in assessing workplace noise, and this is the peak sound pressure during the measurement period – L_{peak}. The maximum sound pressure level (or L_{max}) is a different quantity which is not used in workplace noise assessment – it will always be lower than the required peak value. Some equipment manufacturers apply the term 'peak' to the highest sound pressure occurring during the previous second. In that case the overall peak value will probably be called the 'maximum peak'. It is important when measuring peak sound pressure that the correct quantity is measured.

Because the peak sound pressure normally occurs for a very short period, it is not strictly covered by the definition of the decibel scale using rms sound pressure. Therefore it is sometimes expressed as a simple pressure in pascals. For most purposes it is equally acceptable to express it in decibels (as most sound

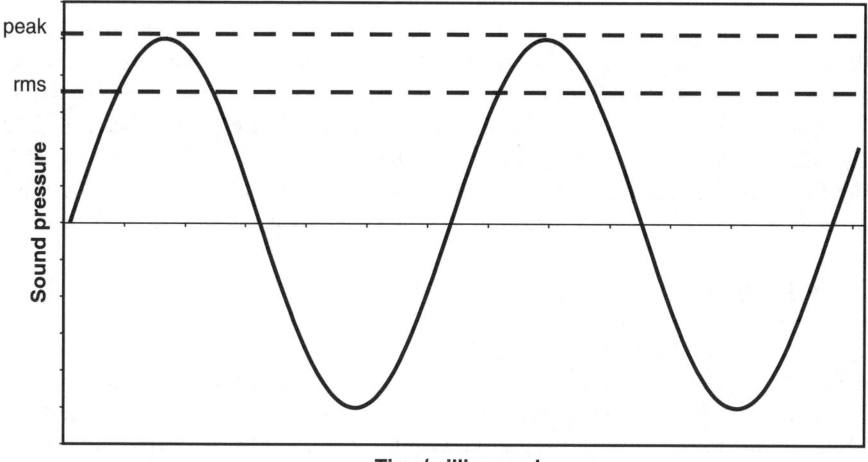

Figure 3.2 Peak and rms levels.

level meters do), and the peak action level under the Noise at Work Regulations can equally be described as 200 Pa or 140 dB.

For a continuous sine wave, the peak level will be 3 dB above the sound pressure level (which in this case will be equal to the L_{eq}). For most real sounds, the difference between L_{peak} and L_{eq} will be greater than this (Figure 3.2). For some impulsive noises, it will be very much greater.

BS EN 61672 includes a precise definition of the peak time constant which is more complete than that in the previous version of the standard. As a result, there may be small differences in peak readings between sound level meters made to BS EN 61672 and those complying with earlier standards.

Frequency weighting

Noise encountered at work normally contains a mixture of different frequencies, and these are normally dealt with by choosing an appropriate frequency weighting when making the measurement. For assessments of daily noise dose this will be the A weighting system, which is available on every sound level meter. The very cheapest sound level meters are capable only of A weighted measurements, so that it is not possible to make peak measurements for comparison with the peak action level. This may be acceptable if it has previously been established that the peak action level will not be exceeded.

The peak action level defined in the original European Directive 86/188/EC and in the Noise at Work Regulations 1989, required the use of a linear weighting. This is available on many sound level meters and implies that all frequencies between 20 and 20 000 Hz are treated equally in assessing the overall level. It was realized after the issue of the regulations that, while linear weightings are very common on sound level meters, no instrument standard specifies in detail the frequency range to be covered, the maximum allowable deviation from a flat response and so on. It was suggested that for many purposes the C weighting would give similar results to a linear measurement and this had the merit of leading to consistent results since it was fully defined in IEC 651, the then current standard for sound level meters. The 1989 Machinery Directive does in fact specify the use of C weighting when measuring noise emissions from machinery.

The results using C weighting can in some cases be significantly different from those using a linear weighting.

The new standard IEC 61672 defines a new frequency weighting, the Z weighting (Figure 3.3). This is essentially the linear weighting but is now put on a sounder footing so that any instrument with a Z weighting should output the same figures in the same situation. C weighting will continue to be available on some sound level meters, and indeed the Physical Agents (Noise) Directive specifies peak action and limit values in terms of the C weighted peak levels. IEC 61672 also allows sound level meter manufacturers to continue to provide a frequency weighting based on the old linear weighting. This is to be defined by the manufacturer, and is to be named 'flat' rather than 'linear'. Sound level meters already having a linear weighting will continue in use for many years, and peak measurements made using linear, flat, or Z weighting will not normally differ very greatly in practice from those using C weighting (the exception to this will be if the peaks occur at the extremes of the frequency range).

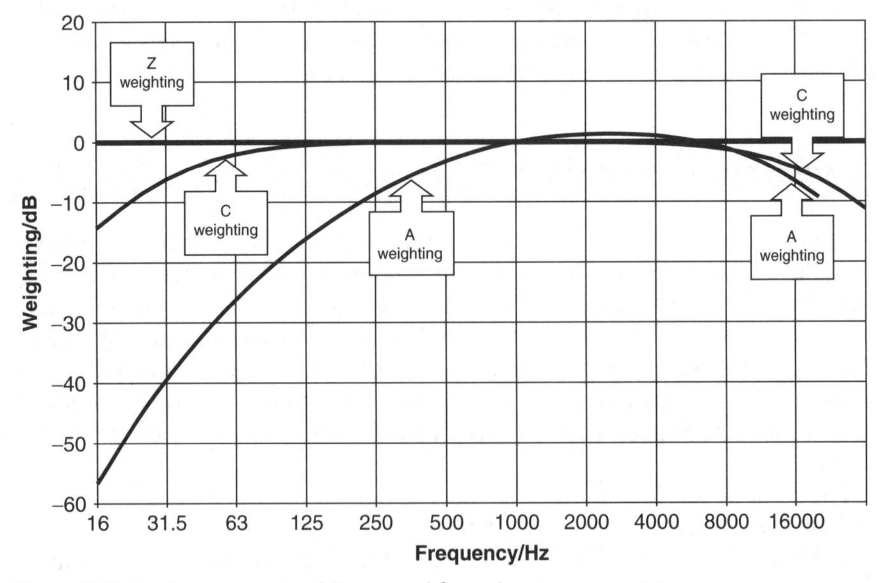

Figure 3.3 The frequency weightings used for noise measurement.

Where possible, though, measurements of peak exposure assessments should use C weighting to ensure easy comparisons with future measurements made in accordance with the Physical Agents (Noise) Directive. Peak measurements using A weighting are definitely wrong.

C weighted measurements are also used for some of the procedures for assessing the effectiveness of hearing protectors. C weighting may also find use in some areas of building acoustics, where the difference between the A weighted and C weighted levels is a simple way of checking the low frequency content in the noise environment.

Frequency analysis

Sometimes more information is required about the noise environment than can be extracted from weighted values which, in some degree, are influenced by the entire frequency range. In this case, the audible range is divided into frequency bands, and normally a measurement is made in each band. The sound level meter may measure simultaneously in all bands (in which case it is called a real-time analyser), but these are expensive. More likely, it will measure each frequency band in turn, controlled either manually or by internal software.

In the workplace, octave bands are the most useful. Each octave band is named after its centre frequency, and the centre frequency of each band has twice the centre frequency of the next lower band. Thus the word octave here has the same meaning as it has in music. Octave band measurements are used for the accurate prediction of the attenuation provided by hearing protectors, and can also be used for a variety of other prediction techniques in noise control. The limiting and centre frequencies of the octave bands are shown in Table 3.1. Octave filters are available as external units for some sound level meters, although increasingly they come – as an option – built into sound level meters.

⅓ octave bands – formed by dividing each octave band into three parts – are used in building acoustic measurements and may be used for more accurate noise control work. Narrower bands such as 1⁄12 octave are used more rarely, particularly to identify prominent tones in a broadband noise.

Table 3.1 Octave bands

Centre frequency	Minimum and maximum frequencies
31.5 Hz	22–45 Hz
63 Hz	45–89 Hz
125 Hz	89–177 Hz
250 Hz	177–354 Hz
500 Hz	354–707 Hz
1 kHz	707–1414 Hz
2 kHz	1414–2828 Hz
4 kHz	2828–5657 Hz
8 kHz	5657–11 313 Hz

Sound level meters

The construction and performance of sound level meters are specified in part 1 of the international standard IEC 61672 (adopted as BS 61672:2003). This replaced two much earlier standards, IEC 60651 and IEC 60804, and these were issued as British standards; most recently as BS EN 60651 and BS EN 60804 but formerly as BS 5969 and BS 6698. Many sound level meters manufactured to the specification of the old standards will be in use for many years to come, and the differences between the old and new standards are, for practical purposes, small.

IEC 651 and 804 divided sound level meters into four categories designated, in order of decreasing accuracy, types 0, 1, 2 and 3. IEC 61672 dispenses with types 0 (difficult to realize in practice) and type 3 (not sufficiently accurate for professional use) and defines only classes 1 and 2, which are essentially the same as the types 1 and 2 in the old standards.

Some sound level meters are still manufactured which do not reach the specification of class 2 (indeed some were previously made which even fell below the old type 3 specification), but they are not suitable for assessing workplace noise. They may be satisfactory, though, for an initial survey to see whether more precise measurements are required. Workplace noise measurements may be carried out with either class 1 or class 2 instruments, but class 1 is preferred, especially when there may be significant low- or high-frequency noise components present. The limits on the accuracy of a class 2 meter are much wider at the extremes of the frequency range.

The allowable measurement tolerances for the two classes of sound level meter depend on the frequency involved. They are illustrated in Figure 3.4. In terms of the accuracy of a frequency weighted measurement, it is often assumed that a measurement with a class 1 sound level meter will have an uncertainty of ± 1 dB. A measurement made with a class 2 sound level meter will carry an uncertainty of about ± 1.5 dB. If a significant proportion of the sound energy involved is either at very high or very low frequencies, then the uncertainties may well be greater than this. Furthermore, these uncertainties only relate to the capabilities of the instrument. It should be remembered that there are other sources of uncertainty when assessing workplace noise exposure, and these other uncertainties may sometimes be considerably greater than those due to the accuracy of the instrumentation.

An integrating sound level meter is one which can average sound levels over a time period to indicate L_{eq}. Most workplace noise measurements today are carried out using an integrating sound level meter. Nonintegrating sound level meters are still made and are significantly cheaper than integrating models. Their use is limited, though, to situations in which noise levels are essentially steady over an extended period.

Figure 3.5 shows the essential stages through which the electronic signal from the microphone must pass before sound level information can appear on the display. The all-important microphone is discussed later in this chapter. Other stages also deserve comment.

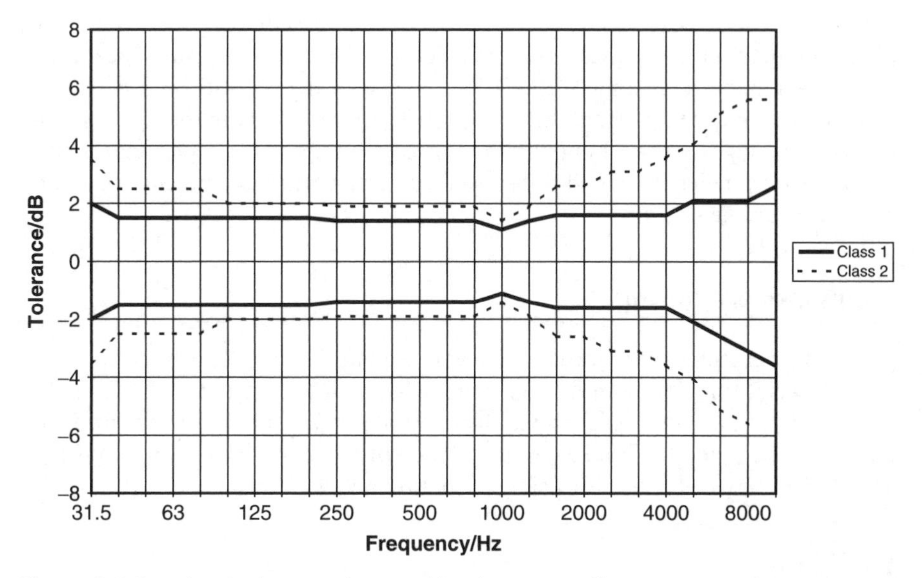

Figure 3.4 Permitted tolerances in sound level meter readings.

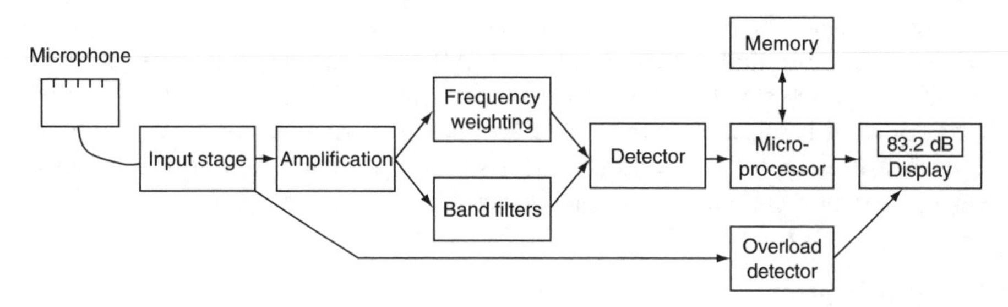

Figure 3.5 Block diagram of a sound level meter.

The input stage

This has the job of converting the microphone signal to a form that standard electronic circuits can easily process. Technically, it converts a high impedance signal to a lower impedance one. High impedance signals are prone to electromagnetic interference which could affect the accuracy of the reading. Many sound level meters offer the facility of removing the microphone and connecting it to the sound level meter via a few metres of cable to facilitate measurements in inaccessible places. If this cable were connected directly to the microphone, then it would be very prone to picking up electromagnetic interference. Instead, the microphone and input stage are removed as a unit and connected to the far end of the extension cable which then carries a low impedance signal.

Modern sound level meters can cover a wide dynamic range (that is the difference between the highest and lowest level which can be measured accurately without changing the range setting). Nevertheless, it may still be necessary to incorporate one or more stages which amplify and/or attenuate the signal so that subsequent stages are working within their capabilities. This is particularly true of sound level meters intended to be used for both workplace and environmental noise measurement.

Frequency weighting

A sound level meter for workplace use will need at the very least to be equipped with A weighting (all sound level meters have this) and either C, Z or flat weighting. Some sound level meters incorporate octave filters so that measurements can be made for use in hearing protector calculations. Older sound level meters may have a facility for using a separate octave filter set.

The detector

This piece of circuitry converts the alternating signal to a direct voltage. Up to this point, headphones connected to the sound level meter would pick up a version of the original noise environment. After the detector, this is replaced by a voltage varying slowly in response to the sound level. It is in this stage that the time constant is selected. If the sound level meter is required to make simultaneous measurements of rms and peak levels (and this facility is useful for noise at work assessments) then two parallel detector circuits are needed.

The integrator

Much of the work, which in an older sound level meter would have been carried out by analogue circuitry, is now performed by a microprocessor. The conversion of sound pressure data into decibel levels, calculation of L_{eq} values, storage of measurement results and formatting of data for a printer are examples of tasks which are normally controlled by a microchip and its ancillary devices.

The overload detector

It is quite possible for the reading on the display to be well within the nominal range of the sound level meter, even though the input stage is presented with a signal which is beyond its handling capacity. This will be particularly true if a lot of high- or low-frequency noise is present as this will largely be removed by the A weighting stage before. It can also be a problem with highly impulsive noises. Measurements under these conditions would not be accurate, even though they seemed to be within the meter's range. It is important therefore that any overload of the input (or any other stage) is detected and independently notified to anyone using the measurement data. If such an overload has occurred, an indication will appear on the display, will be indicated on any printout, and will also be included with any stored data.

Battery condition

A low battery would also lead to invalid data. Modern sound level meters will warn of this and may also give an advance warning if battery voltage is approaching the lowest acceptable level. In many cases they will then switch themselves off so the remaining power can be used to maintain memory and internal clocks. It may not be obvious that these circuits which continue to operate even when the instrument is switched off can be a significant drain on the batteries. If the instrument is stored for a week or two immediately after changing the batteries, it may nevertheless warn of a low battery as soon as it is switched on.

Displays

Sound level meters are no longer manufactured with the old-style analogue displays using a moving needle. This sort of read-out device would not be capable of meeting modern requirements in terms of the dynamic range covered by the display and the resolution required by current standards. Unlike the early digital displays, they had the advantage of showing clearly the direction of any change in the reading. This can be difficult to spot on a digital display. Modern digital displays are very sophisticated, and many include a quasi-analogue indicator. This is normally a moving bar display which mimics the action of an analogue display and complements the digital indication. Most modern sound level meters can calculate many different parameters simultaneously (Table 3.2). A printout can accommodate a large number of figures whereas the digital display normally has space for a relatively small number. Normally these can be selected by the user. The parameters to be indicated for an environmental noise measurement would be different from those which are used in workplace noise assessments.

Table 3.2 Necessary and desirable features of a sound level meter for workplace noise assessments

Necessary	Desirable
Meets at least IEC 61672 class 2	IEC 61672 Class 1
Frequency weightings include A and either C, Z, flat or linear	A and C weighting
L_{eq} and peak measurements	Capable of measuring L_{Aeq} and $L_{C\text{peak}}$ simultaneously
Sound level supplied with a calibrator	Octave band measurements

Microphones for noise measurement

The microphones used for noise measurement need to be extremely accurate and are therefore very expensive. It is largely the quality of the microphone which determines whether a sound level meter will meet the class 1 or class 2 specification (or neither). Two types are used: condenser microphones (Figures 3.6 and 3.7) and electret (or pre-polarized condenser) microphones (Figure 3.8).

A condenser microphone consists of a thin nickel foil stretched across the metal cylinder which forms the body of the microphone. The foil is normally a few microns thick, and a metal back plate is positioned about 20 microns behind

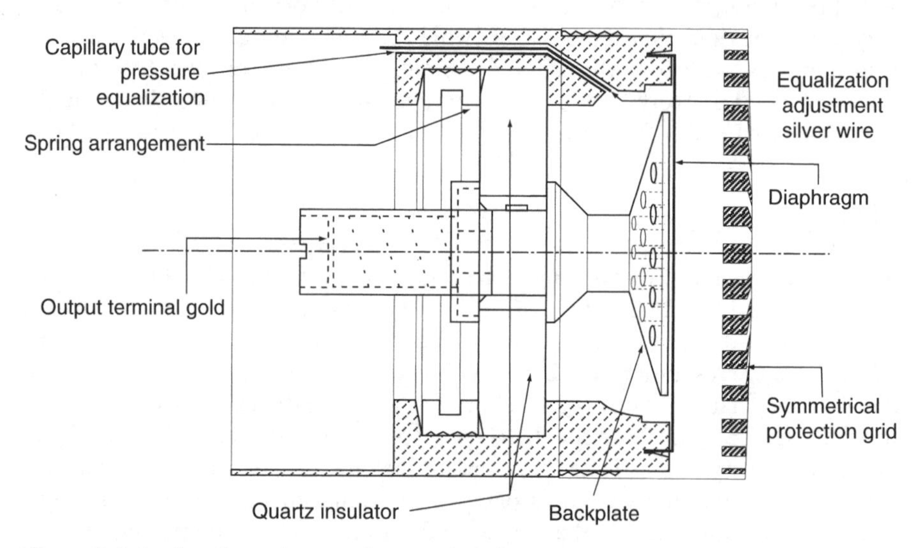

Figure 3.6 Section through a condenser microphone.

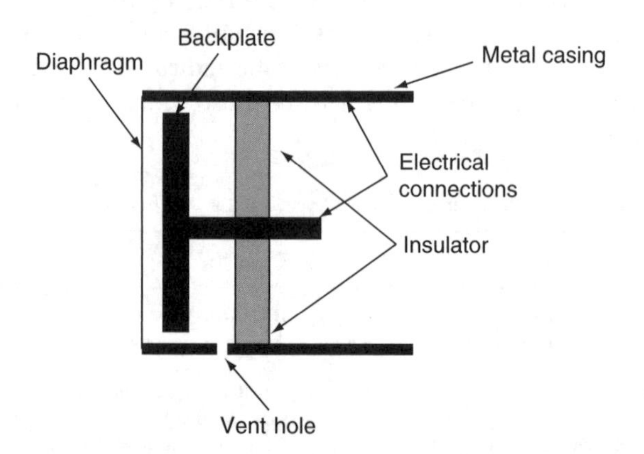

Figure 3.7 Schematic diagram of a condenser microphone.

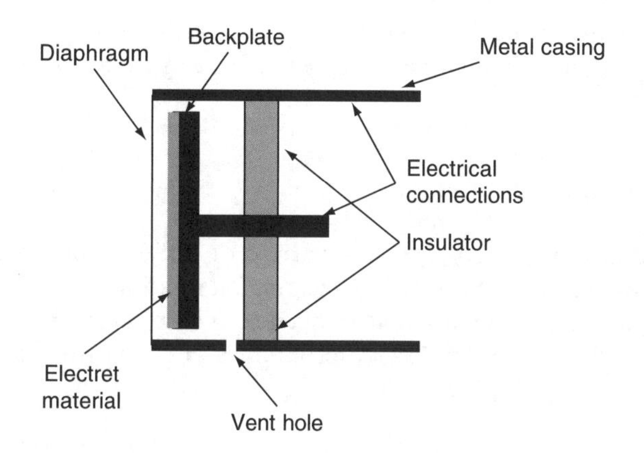

Figure 3.8 Schematic diagram of an electret microphone.

it. Microphones are made to a series of standard physical dimensions named for their diameter in imperial units, the ½ inch microphone being the most common.

A section of the sound level meter generates a high voltage – normally 200 volts – which is used to place a positive charge on the diaphragm and a negative one on the back plate. Pressure fluctuations due to the sound wave will cause the diaphragm to be alternatively pushed towards, and pulled away from, the back plate. The diaphragm and back plate together form a capacitor – a common electronic component whose ability to store charge – its capacitance – is proportional to the separation of two plates. The effect of moving one plate backwards and forwards is to generate a small alternating voltage which constitutes the output from the microphone.

There are other versions of the electret microphone, and indeed some of the very cheapest microphones manufactured – such as those built into portable tape recorders – use the same principle. The version used for noise measurement is constructed in a very similar way to the standard condenser microphone, as shown in Figure 3.8. The difference here is that the back plate is coated with a layer of electret material, which has the property of separating electric charge. As a result, the back plate will act as if it carried a charge, and the diaphragm carried an opposite one. The polarizing voltage is no longer needed and the bulk and complexity of the sound level meter is correspondingly reduced. On the other hand, the microphone is slightly more difficult to manufacture. The accuracy achievable with an electret microphone is slightly lower than that which is possible with a condenser microphone, but nevertheless high quality electret microphones are up to the standards required for class 1 sound level meters.

Class 2 sound level meters nearly always have an electret microphone since the overall manufacturing costs are lower. It may not have the standard physical dimensions which are universal among class 1 instruments. Even though the physical dimensions are the same, it may not be possible to interchange microphones between sound level meters. Consideration needs to be given to the characteristics of the microphones involved before any exchange of microphones is made.

Frequency response

Microphones are manufactured for a variety of different purposes. Most are intended for use in sound level meters meeting international standards, in which case their ability to handle different frequencies will be such that they meet the tolerances specified in IEC 61672. Microphones intended for more specialized purposes may not be suitable.

Sensitivity

The output level from different microphones exposed to the same sound field may be different. The normal calibration facilities allow for small adjustments, but different microphone types may have sensitivities differing by a greater amount than the available adjustment.

Directional response

The frequency response of a microphone depends on the sound field to which it is responding. Sound level meters may be fitted with either a free-field microphone – designed to have flat frequency response when responding to sound waves from in front – or a random incidence microphone – designed to have a flat frequency response in a reverberant field, when sound energy arrives from all directions. In many workplaces the sound field is reverberant, but ISO standards specify a free-field microphone, so a sound level meter used for workplace noise assessment must be fitted with a free field microphone, whether it is to be used indoors or outdoors. Modern sound level meters may be able to correct electronically for the microphone type, and once again it is important that free field incidence is selected.

Polarization

An electret microphone will not be capable of accurate measurements if fitted to a sound level meter which supplies a polarizing voltage. Conversely, a condenser microphone cannot be used without a polarizing voltage. Once again, some sound level meters can be set up to work with either type.

When a microphone is exposed to air currents, noise will be generated close to the diaphragm as a result of air currents through the protective grid. Even at low air velocities, this noise can be significant and will lead to false results. To avoid this, a wind shield is used, which consists of a spherical piece of acoustic foam with a diameter of 70–100 mm and a hole in which to insert the microphone. These are made to a precise specification – DIY copies would not be acceptable – and do the same job as the furry sausages that TV reporters thrust in interviewees' faces. It is good practice to use a windshield outdoors, even though sometimes it is not necessary. Indoors, it is not normally necessary to use one, but in some environments it can be useful to protect the microphone from dust and physical contact.

Dosemeters

A personal dosemeter (Figures 3.9 and 3.10) is a small integrating sound level meter with a cable-mounted microphone which can be fitted to an employee to record sound exposure levels during an extended period such as a whole shift.

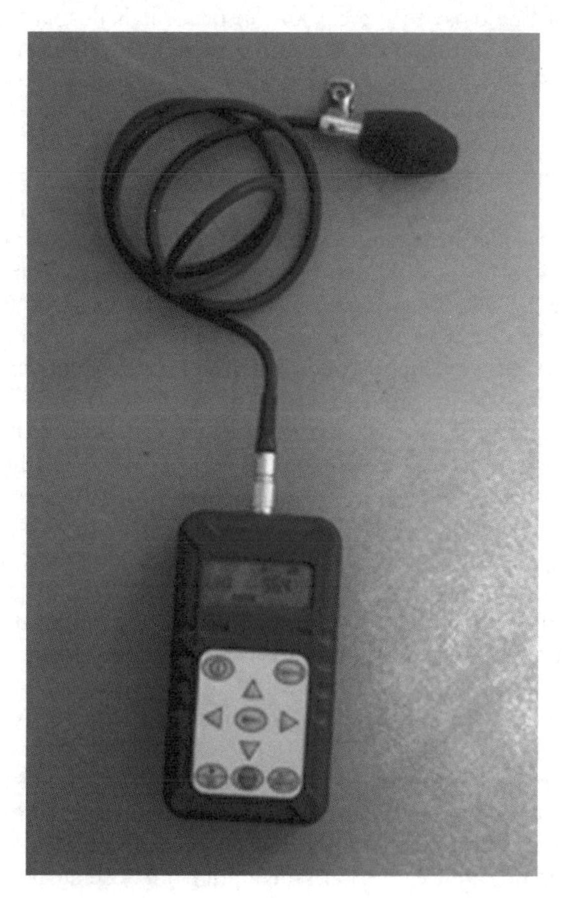

Figure 3.9 A personal dosemeter.

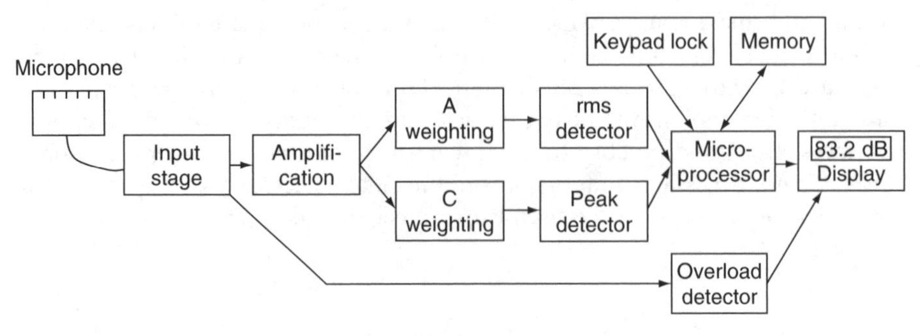

Figure 3.10 Block diagram of a personal dosemeter.

They are equally known as dosimeters and personal sound exposure meters – the latter term is the one which normally appears in standards and official guidance.

A dosemeter can be put in the wearer's pocket or clipped to a belt. The microphone must be close to the ear and in practice is normally clipped to the collar or to a hard hat if one is worn. Normally this means that the microphone is closer than ideal to the head, body or clothing, and this limits the accuracy which is possible. A badly placed microphone may generate spurious readings when it brushes against clothing.

Because the dosemeter operates most of the time without direct supervision, there is an obvious danger that the readings will be tampered with. To some extent this can be guarded against, and a number of systems are in use. For example the controls of a dosemeter can be locked either by an external computer, or by a sequence of button presses which will not be obvious to the wearer. This leaves the possibility of the microphone being exposed deliberately to false levels. It could either be taken off and left in a cupboard, or conversely put next to a loud noise source.

Early dosemeters represented a considerable triumph of electronic miniaturization, and methods of accessing the reading were sometimes rather arcane. Outputs were normally expressed as a percentage dose, with a 100 per cent dose being equivalent to the then advisory exposure limit: an $L_{EP,d}$ of 90 dB. Electronics has advanced swiftly, and it is not difficult for a small device to output its measurement results in a number of different formats. Nevertheless, the percentage dose is sometimes still encountered. Since the advent of the Noise at Work Regulations this is a potential area for confusion since it needs to be specified which action level is being regarded as representing a 100 per cent dose.

Most dosemeters offer a direct $L_{EP,d}$ reading (as do some sound level meters intended for workplace use). This $L_{EP,d}$ value will only be correct if the dosemeter has been worn for one complete shift. If it has deliberately been used over part of a shift, or even if for practical reasons the dosemeter was fitted after the start of work and collected before the end, the $L_{EP,d}$ value will be wrong.

Chapter 5 contains advice on processing data from dosemeters and sound level meters to calculate correct daily noise doses.

Calibration

It needs to be demonstrated when carrying out a workplace noise assessment that the equipment used was capable of making accurate measurements. There are three separate processes which can be described as calibration:

- Type approval
- Field calibration
- Periodic verification.

Type approval

Sometimes called pattern evaluation, this is carried out by an independent laboratory on the early production samples of a new type of measuring instrument. It normally involves detailed tests to check that every aspect of the claimed specification is in fact met by the instrument. In the UK it is not a requirement to have these tests carried out. In Germany, though, an independent test report is necessary to verify the claims that are being made. Since the market for noise measuring equipment is an international one, most sound level meters on the market are in practice approved in this way.

Field calibration

Field calibration is familiar to most of those who have carried out noise measurements. It involves the use of the calibrator normally supplied with a sound level meter (Figure 3.11). The international standard IEC 60942 specifies the characteristics of these calibrators, which are divided, like sound level meters into different classes. A class 1 sound level meter must be calibrated with a class 1 calibrator. The calibrator needs to be correctly fitted to the microphone, when it will generate a repeatable sound level in the cavity between the microphone and the calibrator's own transducer. Normally this is a pure tone with a frequency of 1 kHz. The sound pressure level generated varies between instruments.

Although most calibrators intended for use with ½ inch microphones can be fitted to any ½ inch microphone, the level generated may be slightly different if not used with the particular microphone type for which it was designed. Manufacturers can normally advise on the expected output level for a particular combination of calibrator and microphone.

A simple sound level meter may need to be adjusted with a small screwdriver so that the reading is the same as is specified for the calibrator with the microphone in use. More sophisticated instruments will have an automatic calibration procedure which requires only that the calibration level is entered and a button pressed to initiate the calibration. Dosemeters will similarly have a calibration programme. Many modern noise measuring instruments are

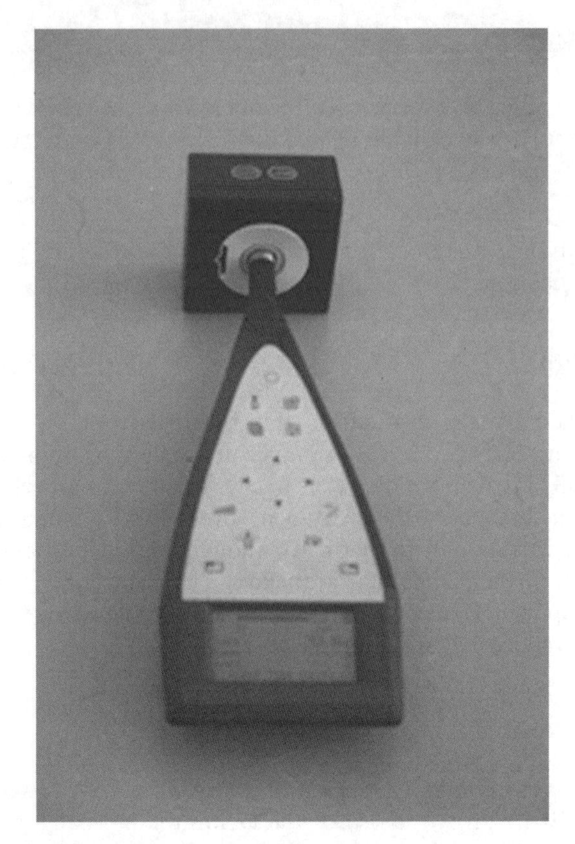

Figure 3.11 Field calibration of a sound level meter.

in any case extremely stable, and it will not normally be necessary to adjust the reading. Nevertheless, calibration is always carried out before and after measurement as confirmation that the instrument is operating correctly. In the case of a long series of measurements – occupying a whole day, for example – one or more additional calibrations may be carried out between measurements.

If field calibration is carried out as described, there are still two possible problems with the reliability of subsequent measurements:

1. If the calibrator is not performing to specification, then all the measurements made will be wrong.
2. The calibrator tests the accuracy of the sound level meter when measuring a 1 kHz pure tone at one particular level. Real noise measurements will be of a mixture of frequencies at various levels fluctuating with time. It would be quite possible for a faulty sound level meter to respond correctly to the calibrator but to make inaccurate measurements in the field.

Periodic verification

To check a much wider range of functions than are covered during a field calibration, a sound level meter can be sent to an independent laboratory which will carry out the tests specified in the two parts of BS 7580 (part 1 for class 1 sound level meters, part 2 for class 2 meters; both parts will eventually be replaced by part 2 of IEC 61672). The sound level meter and its calibrator are sent for verification together. The laboratory will test the response of the sound level meter at a range of frequencies and at different sound levels. It will check the operation of the time constants and of the averaging process used in L_{eq} measurement. It checks on the instrument's ability to accept signals with a moderate crest factor, and it also confirms that the overload indication operates correctly.

The tests prescribed in BS7580 do not test every function available on most sound level meters. More importantly, the verification certificate does not guarantee the accuracy of the sound level meter for any period in the future, since the test laboratory has little control over what happens to the instrument even while it is in transit back to the customer. The test certificate only confirms that the sound level meter met a particular specification at the time the tests were carried out.

The period between verifications can be contentious. A number of standards, and the Health and Safety Executive's guidance on workplace noise assessments (HSE, 1998), specify a maximum of 2 years between verifications. This is an absolute maximum which may be an appropriate period in the case of an instrument used infrequently and under controlled conditions. Often, though, it will be necessary to verify more frequently, and some of the factors affecting this are shown in Table 3.3. How often to send a sound level meter for verification is a decision which should be made by the person responsible for the reliability of the measurement results. Depending on the type of organization involved, this could be the health and safety manager or a quality manager. Many calibration laboratories will recommend annual verification – frequently using phrases such as 'out of calibration' if this period is exceeded – and a manufacturer may even arrange for instrument printouts to specify a re-verification date. However, these recommendations are based on the commercial interests of the verification laboratory and do not override the professional judgement of the person responsible for the accuracy of the sound level measurements to be made.

Table 3.3 Some factors affecting the period between verifications

Tending towards less frequent verifications	Tending towards more frequent verifications
Infrequent SLM use	Frequent use
Used by one operator	Used by a number of operators
Used only in nonextreme environments	Exposed to dust, corrosive or humid atmospheres or extremes of temperature
Used mainly for preliminary surveys before commissioning more precise measurements	Measurements may form the basis for important decisions, or may in future be required as evidence in court cases

The aim of arranging an external verification is to ensure that those concerned – that is those carrying out workplace noise assessments and those with a duty to act on the results – have reasonable confidence that instrument deficiencies have not made a major contribution to the uncertainty which will exist in every exposure assessment.

Development of workplace noise control in the UK

Early work

Early hunter-gatherers may have been exposed to high noise level from time to time, and medieval blacksmiths and masons probably risked hearing damage as a result of their trades, but it was not until the industrial revolution that large numbers of workers were exposed to levels consistently high enough to cause widespread damage to their hearing. It was probably recognized early in the nineteenth century that the noise exposure in certain trades and industries commonly caused permanent hearing impairment, but few employers regarded the problem as their concern. Many affected workers may have accepted deafness as being a necessary consequence of their occupation, and preferable to unemployment. Certainly deafness was less serious than many of the other industrial illnesses which were prevalent at that time, and to which the newly formed UK Factory Inspectorate turned their attention.

Thomas Barr, a Glasgow aural surgeon, made a detailed study of industrial deafness in the 1880s (Barr, 1886). He ventured into shipyards and foundries to gain first hand experience of the noise levels to which his patients were exposed, and he saw the need to compare their hearing ability with a control group of postmen who had not been exposed to high noise levels at work. In the absence of anything approaching a modern audiometer, he used a variety of methods to gauge his patients' hearing. Most famously, he measured patients' hearing by holding his pocket watch at various distances from their ears to arrive at an assessment of their hearing ability which he expressed in inches. The paper he delivered to the Glasgow Philosophical Society in 1886 includes an assessment of the effect of noise exposure on hearing ability at different frequencies, as well as a discussion of suitable forms of hearing protector. It even has advice for clergymen working in industrial cities, whose deaf congregations might miss out on necessary moral guidance as a result of their disability!

Noise and the worker

It was more than 60 years after Thomas Barr's work that a UK Government took an active role in the prevention of industrial deafness. The Wilson report, published in 1963, had looked at noise in a wide variety of contexts, and its recommendations on the control of industrial noise exposure led to the publication of the booklet *Noise and the Worker* by the Ministry of Labour later that year. The relationship between sound pressure level, frequency and exposure time was not sufficiently well established to include a definitive method for the measurement and assessment of noise exposure even if suitable equipment had been widely available. Even so, the booklet today contains much that is familiar to those working in the field of workplace noise management, and a great deal of good advice about reducing noise levels in industry.

It has always been tempting to see noise exposure as a feature of heavy industry. It is interesting to observe that in the late 1960s, a number of local authorities began to be concerned about hearing damage to young people attending discos and other establishments where amplified music was played. Eventually it became apparent that their concern was somewhat misplaced as it is rare that customers attend establishments such as this consistently enough to accumulate significant hearing loss. What few people spotted at the time was that those who worked in these premises were exposed to high noise levels for much greater periods, and were much more likely to suffer hearing damage as a result.

The 1972 code of practice

1972 saw the publication of the government's *Code of Practice for Reducing the Exposure of Employed Persons to Noise*. Although one or two groups of employees were protected before this by legislation such as the 1963 Woodworking Machines Regulations, this was the first attempt to reduce the noise exposure of all workers. It was an advisory limit only, but with the advent of the Health and Safety at Work Act in 1974, there was a possibility of taking action to limit noise exposure on the ground that an employer allowing exposure to rise above this level was not complying with the duty to ensure, so far as was reasonably practicable, the health, safety and welfare of employees.

Noise and the Worker had used a rather complicated frequency-dependent assessment of noise exposure, but the 1972 code replaced this with an index of noise exposure which was broadly equivalent to the modern $L_{AEP,d}$ although its use was limited by the measuring equipment available at that time. A recommended $L_{AEP,d}$ limit of 90 dB was adopted. An A weighted level of 135 dB (using fast time constant), and, in the case of 'impulse noise', an instantaneous sound pressure of 150 dB were additionally recommended as 'overriding limits'. By modern standards, the measurements required to establish compliance with these overriding conditions were not adequately specified. It is clear, though, that neither is fully comparable to the peak action level of the Noise at Work Regulations.

The 1974 Health and Safety at Work Act marked a turning point in public and private attitudes to workplace health and safety, and work was soon under way to develop a set of statutory limits on noise exposure. It was considered that first it was necessary to establish more conclusively the quantitative relationship between noise exposure and hearing damage, and research was carried out to improve the state of knowledge. Advances in electronics were also making more sophisticated noise measuring instrumentation available, including sound level meters which could directly measure L_{eq} (which previously had to be calculated from a series of sample sound pressure level measurements), and portable digital dosemeters which gave a direct (but not necessarily straightforward) readout of the wearer's accumulated daily noise dose.

European Directive 86/188 and NAWR

By 1980, the work on occupational noise exposure regulations was well advanced, but at this point it became clear that there would soon be European legislation on the subject, and efforts were diverted towards influencing the shape of the European legislation rather than to developing UK regulations. After much debate about appropriate levels at which to set noise exposure limits, the 1986 Noise Directive was issued which passed into UK legislation as the Noise at Work Regulations 1989. Some member states had proposed a single action level at an $L_{AEP,d}$ of 85 dB – in line with their own domestic legislation. The then UK Government had fought hard against a limit of 85 dB, which it believed would impose an unreasonable burden on industry. As a result of the debate on an appropriate limit, two action levels emerged based on the equivalent A weighted 8-h exposure $L_{AEP,d}$. Employers had to comply with various duties when the noise exposure of any employee exceeded one or other of these action levels.

The peak action level was set at a sound pressure of 200 pascals, to be measured 'unweighted'. The use of the pascal to specify peak levels is discussed in Chapter 3, and in practical terms this is more usefully described as 140 dB. There is a problem with the need for an 'unweighted' measurement. Originally intended to refer to a linear weighted measurement, it was later realized that no rigorous definition of the linear weighting exists, so that different measuring instruments might be expected to record significantly different peak levels when exposed to the same noise environment. As a result, it is common to use the C weighting (which is rigorously defined in IEC 61672 and its precursor standards) as an alternative to linear weighting. However the uncertainty introduced by using C as opposed to linear weighting is likely to be greater than the difference due to different interpretations of linear weighting.

The Physical Agents Directive

The Noise at Work Regulations became law on 1 January 1990. The dust had hardly settled before the European Union was working on a more ambitious Physical Agents Directive, which would establish a common framework for the regulation of noise, vibration, electromagnetic radiation and optical radiation. The intention so far as noise was concerned was to improve the definitions relating to the peak action level, and to reduce the levels at which the first and second action levels were set. The proposed directive was probably too ambitious. As well as disagreement about the desirability of reducing action levels (as opposed to improving the enforcement of the existing regulation) there were considerable doubts whether it was appropriate to try to force very different agents into the same regulatory framework. From 1994 to 1999 no progress was made. At this point the decision was made to move ahead with a directive to limit exposure to vibration. Action to limit vibration exposure was thought to be urgent as it was not as yet covered by European legislation. Shortly afterwards work started on a new directive to improve the regulation of workplace noise exposure.

The Physical Agents (Noise) Directive passed through its final stages and was published in February 2003. Following the normal implementation period of 3 years, member states will enact legislation to give it effect at the beginning of 2006.

Apart from the changes to the action levels, the Physical Agents (Noise) Directive differs in a number of ways from the Noise at Work Regulations:

1. The duty to carry out a noise exposure assessment is more general than before. On the other hand, it is no longer assumed that an assessment will normally involve measurements.
2. There are some changes in the terminology. $L_{AEP,d}$ is now called $L_{EX,8\,\text{hours}}$. Action levels are now called exposure action values.
3. The old peak action level becomes a new peak exposure limit value. Two new peak exposure action values are defined.
4. The frequency weighting for peak measurements is clarified. It is now to be measured using C weighting.
5. An explicit provision is made for averaging varying daily exposures over a week in cases where exposure from day to day varies considerably.
6. Limit values are now prescribed for both $L_{AEP,d}$ and $L_{C\text{peak}}$. These are to be assessed, unlike the action values, after taking into account the effect of any hearing protection worn. They are levels above which employees must not be exposed, and if exposure above these levels which does take place, then

employees have a duty immediately to reduce exposure below them and to take steps to prevent a recurrence.

The new action and limit values are listed in Table 4.1. Table 4.2 compares the duties under the 1986 Noise at Work Regulations and the Physical Agents (Noise) Directive. The regulations which eventually implement the latter into UK law may differ in detail from the directive itself. Appendix B also details the duties under both pieces of legislation.

Table 4.1 Exposure action and limit values under the Physical Agents (Noise) Directive

	$L_{AEP,d}$	L_{Cpeak}
Lower exposure action value	80 dB	135 dB or 112 Pa
Upper exposure action value	85 dB	137 dB or 140 Pa
Exposure limit value	87 dB	140 dB or 200 Pa

Table 4.2 Employer duties under the Noise at Work Regulations 1986 and the Physical Agents (Noise) Directive

Duty	Noise at Work Regulations 1986	Physical Agents Directive
Reduce the risk of hearing damage to the lowest limit reasonably practicable	At any exposure level	At any exposure level
Arrange for a noise exposure assessment to be carried out by a competent person	If it appears likely that the exposure of an employee will exceed the first or peak action levels	An assessment (not necessarily involving measurements) must be made of the noise exposure of every employee
Keep records of the assessment until a new one is carried out		
Provide information and training to employees about risks to hearing, how to minimize those risks, etc.	Above the first action level or the peak action level	Above the lower exposure action values
Carry out hearing tests	Not explicitly required	Above the upper action values
Provide hearing protection to employees who ask for it	Above the first action level	Above the lower exposure action values
Maintain noise reduction equipment and hearing protection	When provided under the regulations	When provided under the regulations
Reduce exposure to noise by means other than hearing protection	Above the second or peak action levels	Above the upper exposure action values
Provide hearing protection to employees and ensure that it is used	Above the second action level or the peak action level	Above the upper exposure action values
Mark hearing protection zones, and ensure that all those entering them wear hearing protection	Above the second action level	Above the upper exposure action values
Ensure that employee exposure does not exceed ...	Not specified	... the exposure limit values

Setting the agenda

Much attention is focused when changes to health and safety regulation are proposed on the exact limits and action levels which should be adopted. Organizations representing employers normally oppose lower limits or stricter regulations because of the extra costs they will impose, reducing the profits of some companies and possibly forcing others out of business with loss of jobs in the company and its suppliers. Trade unions traditionally call for lower limits and tighter regulation to protect their members' health. Professional bodies involved in the management and enforcement of health and safety also have a vested interest in increasing the volume and significance of their members' role. Meanwhile, some of those whose health the debate is supposedly about take a cavalier attitude to the protective measures their representatives are promoting.

In the political sphere it is normal for both ministers and opposition leaders to claim that their position rests on a solid scientific base, while their opponents are taking action based on 'political' considerations (which by implication are short-term and motivated by considerations of personal gain).

It is easy to overestimate the role of scientific investigation in any such issue, and workplace noise exposure is no exception. At best, science can establish the relationship between the causative factor (exposure to high levels of noise) and the human effect (noise-induced hearing loss). In doing so, it will be necessary to establish how both the cause and the effect are to be quantified, both in terms of what quantity is to be measured, and what protocols are to be used when making the measurement. If reproducible measurements cannot be made, then it is very difficult to enforce any exposure limit.

In the case of noise, it has been established that the A weighted personal daily exposure $L_{AEP,d}$ is the most useful measure of noise dose, and that the sum of the hearing loss measured at 4000, 6000 and 8000 Hz can be used to quantify the effects of noise-induced hearing loss. Figure 4.1 shows a relationship between exposure and health effect using these measures. The meaning of this particular graph is not explained here (see BS 5330:1976 for more details), and it is open to challenge on a variety of grounds. The point here, though, is that the establishment of this kind of relationship is as far as science can take us. Deciding how much hearing loss is acceptable as a result of workplace exposure and, crucially, what economic costs are acceptable in order to reduce health effects of noise exposure must be a political decision. This means not that it should be left to politicians but that no one person's opinion is worth more than anyone else's. A worker in a noisy factory will have a valid view on this, as will his or her relatives, the neighbours who might be exposed to high volumes from their TV set and those involved in the health care of occupational deafness sufferers.

At one extreme, it has been argued that no limits should be permitted by law which leave open the possibility of any employee suffering hearing damage. At the opposite extreme is the viewpoint that individuals should be free to decide whether or not to expose themselves to hearing damage and all that is required is

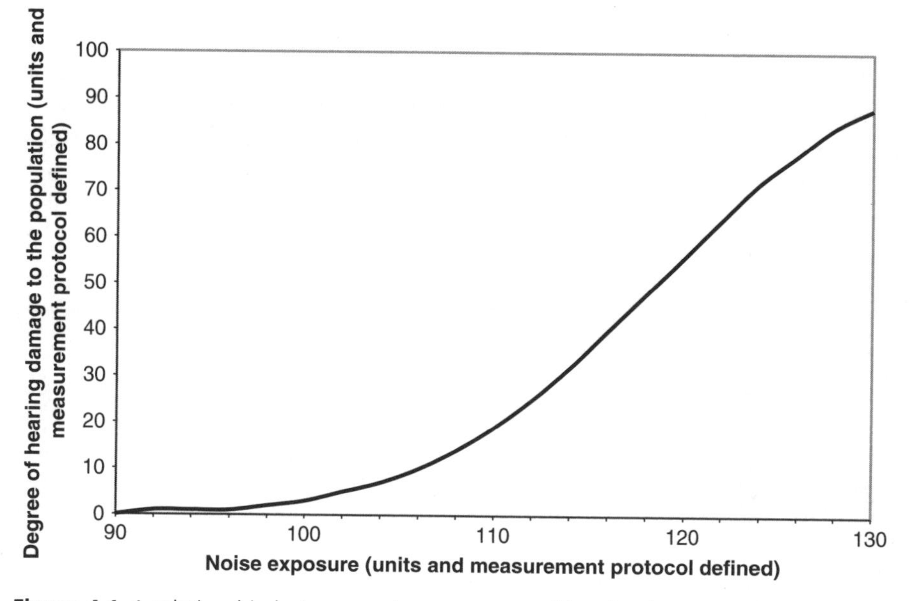

Figure 4.1 A relationship between noise exposure and hearing loss.

that they should be fully informed of the risks so that they can make an informed decision. This last position has several difficulties, one of which is that it is probably easier to institute hearing conservation measures than to make sure that every individual is fully informed of the risks they may be taking. It is now widely agreed that individuals should be protected by statutory exposure limits, the setting of which will involve weighing up the various risks and benefits involved. The aim of preventing any occupational hearing loss completely is a difficult one to achieve economically, and there is in any case no agreement yet on a level of exposure which will not damage the hearing of even the most sensitive individual.

As societies become more prosperous and life expectancy is increased, more people will expect to be able to enjoy full use of their hearing into retirement. At the same time, the existence of limits and action levels encourages engineers and managers to develop quieter machinery and more sophisticated noise control techniques. As a result it will become easier to enforce still lower action levels. It is probable therefore that there will be a gradual downward pressure on limits and action levels.

The sometimes furious arguments about where exposure limits and action levels should be established, sometimes obscures the remarkable consensus that now exists about many aspects of the control of workplace noise exposure. Within the European Union there is little debate about a number of key issues:

- The duty on employers to arrange noise exposure assessments if it appears that a particular action level is likely to be exceeded.

- The quantities to be measured and appropriate procedures for carrying out the measurements for these assessments.
- The need for a variety of actions to be taken if exposure rises above various action levels: noise control at source, health surveillance, use of hearing protection, storage of assessment records and arrangements for periodic reassessment.

Calculating noise doses and limits

Introduction

In this chapter the various calculations which may be required during or after a workplace noise assessment are collected together. In each case a worked example is included, and further examples can be found in the appendix. The various calculations which may be required are:

- Personal daily dose calculations
  - From one or more subperiods
  - From single event data
- Weekly average dose
- Conversion of $L_{EP,d}$ data to and from other systems for expressing noise exposure
  - Noise exposure, E_A
  - Percentage dose
- Ranking various time periods in order of their contribution to the total noise dose
- Setting of an exposure time limit
  - Involving a single noise level
  - Assuming previous noise exposure.

Working out the dose

The simplest sort of calculation to be done as a result of workplace noise measurements is to work out the personal daily dose $L_{AEP,d}$ of an employee. This can be done directly from a knowledge of the levels to which the employee was exposed during the working day, along with the period for which exposure to each level took place.

$$L_{EP,d} = 10 \times \log\left(\frac{\left(t_1 \times 10^{\frac{L_1}{10}} + t_2 \times 10^{\frac{L_2}{10}} + \ldots\right)}{8}\right) \quad (5.1)$$

Here, L_1, L_2 are the various levels to which the employee is exposed during different periods; t_1, t_2 are the corresponding exposure times.

The number 8 in the denominator represents an assumed working day of 8 h. This standard working day is used irrespective of how many hours the employee actually works. It assumes that all time periods are measured in hours. If for some reason it is decided to measure all times in minutes, then the 8 must be replaced by 480, the number of minutes in 8 h.

It is assumed above that there are two (or more) periods during which sound pressure levels are essentially fixed. In practice, though, this is not necessary. If noise levels fluctuate, then L_1, L_2, etc., can be the L_{eq}s for each subperiod over which it was convenient to make measurements. Two levels and two corresponding time periods are shown in the equation above, but in practice there is no limit to the number of subperiods into which the working day is divided.

Occasionally the level is constant during the whole of a shift. Alternatively a fluctuating sound pressure level may have been measured during the whole of a shift. In this case, the measured L_{eq} can be converted directly into the daily dose:

$$L_{EP,d} = L_{eq} + 10 \times \log\left(\frac{t_m}{8}\right) \quad (5.2)$$

Examples

1. Calculate the personal daily dose, $L_{EP,d}$ which results when an employee is exposed to a level of 92 dB for 2 h, followed by a level of 84 dB for 5 h.

$$L_{EP,d} = 10 \times \log\left(\frac{\left(t_1 \times 10^{\frac{L_1}{10}} + t_2 \times 10^{\frac{L_2}{10}}\right)}{8}\right) = 10 \log\left(\frac{\left(2 \times 10^{\frac{92}{10}} + 5 \times 10^{\frac{84}{10}}\right)}{8}\right)$$

$$= 87.4 \approx 87\,\text{dB}$$

2. An employee is exposed to noise levels during the periods of the working day detailed in the table. Work out this employee's daily exposure $L_{AEP,d}$.

Task	Period	L_{eq}
Thicknesser	2 h	102 dB
Frame saw	30 min	96 dB
Band saw	1 h	85 dB
Router	3 h	93 dB
Panel saw	90 min	90 dB

$$L_{EP,d} = 10 \times \log\left(\frac{\left(t_1 \times 10^{\frac{L_1}{10}} + t_2 \times 10^{\frac{L_2}{10}} + t_3 \times 10^{\frac{L_3}{10}} + t_4 \times 10^{\frac{L_4}{10}} + t_5 \times 10^{\frac{L_5}{10}}\right)}{8}\right)$$

$$= 10 \times \log\left(\frac{\left(2 \times 10^{\frac{102}{10}} + 0.5 \times 10^{\frac{96}{10}} + 1 \times 10^{\frac{85}{10}} + 3 \times 10^{\frac{93}{10}} + 1.5 \times 10^{\frac{90}{10}}\right)}{8}\right)$$

$$= 97.1\,\text{dB}$$

3. A glass collector is exposed to an L_{Aeq} of 99 dB during a 3 h shift. What is this person's daily dose $L_{AEP,d}$?

$$L_{AEP,d} = L_{Aeq} + 10 \times \log\left(\frac{t_m}{8}\right) = 99 + 10 \times \log\left(\frac{3}{8}\right) = 94.7\,\text{dB}$$

Calculating $L_{AEP,d}$ from L_{AE}

L_{AE} can be measured directly on many sound level meters, or it can be calculated from the L_{eq} during the event, plus the duration of the event:

$$L_{AE} = L_{eq} + 10 \times \log t \tag{5.3}$$

where t is the duration of the event as measured.

The daily exposure can then be calculated from this L_{AE} and the number of similar events to take place during the shift:

$$L_{EP,d} = L_{AE} - 10 \times \log 28\,800 + 10 \times \log n \tag{5.4}$$

or

$$L_{EP,d} = L_{AE} + 10 \times \log n - 44.6 \tag{5.5}$$

(28 800 is the number of seconds in 8 h, and $10 \times \log(28\,800)$ is 44.6).

Example

An employee is exposed to noise from low-flying aircraft 15 times during a shift. Measurements on a typical overflight show an L_{Aeq} of 93 dB, measured over 60 s. Calculate this employee's daily exposure.

$$L_{AE} = L_{eq} + 10 \times \log t = 93 + 10 \times \log 60 = 110.8\,\text{dB}$$

$$L_{AEP,d} = L_{AE} + 10 \times \log n - 44.6 = 110.8 + 10 \times \log 15 - 44.6 = 80\,\text{dB}$$

Using the HSE Chart to work out a noise dose

Figure 5.1 shows how the chart published by the Health and Safety Executive in Reducing Noise at Work (1998) can be used to calculate noise doses. This chart avoids carrying out any calculations, and is often found to be useful by those who need to assess noise doses occasionally. The measured noise level at the workstation is marked on the left hand column and the exposure time on the right hand column. A line drawn to join these two points will then intersect the middle column at a point corresponding to the daily noise dose, $L_{EP,d}$. $L_{EP,d}$ itself can be directly read out, in this simple case, on the left hand side of the middle column. In this example, the $L_{EP,d}$ is calculated which results for the glass collector in an earlier example who is exposed to a level of 99 dB for 3 h. The equivalent 8-h exposure is shown to be just under 95 dB.

The right hand side of this column can be used to calculate the noise dose in more complicated situations where an individual is exposed to more than one level in the course of a working day. In this case lines are drawn to join each identified exposure level to its corresponding duration. The value of the fractional exposure is read from the right hand side of the middle column. The various fractional exposures are totalled, and finally the middle column is consulted again to find the value of $L_{EP,d}$ which corresponds to this total.

The weekly average dose

In some cases under the Noise at Work Regulations, it is permissible to average the daily personal noise dose over a week. The Physical Agents (Noise) Directive makes this procedure more explicit. The averaging calculation is as follows:

$$L_{EP,w} = 10 \times \log\left(\frac{\left(10^{\frac{L_{EP,d1}}{10}} + 10^{\frac{L_{EP,d2}}{10}} + 10^{\frac{L_{EP,d3}}{10}} + 10^{\frac{L_{EP,d4}}{10}} + 10^{\frac{L_{EP,d5}}{10}}\right)}{5}\right) \tag{5.6}$$

where $L_{EP,w}$ is the daily exposure averaged over the week; and $L_{EP,d1}$, $L_{EP,d2}$, etc., are the $L_{EP,d}$ values for each day.

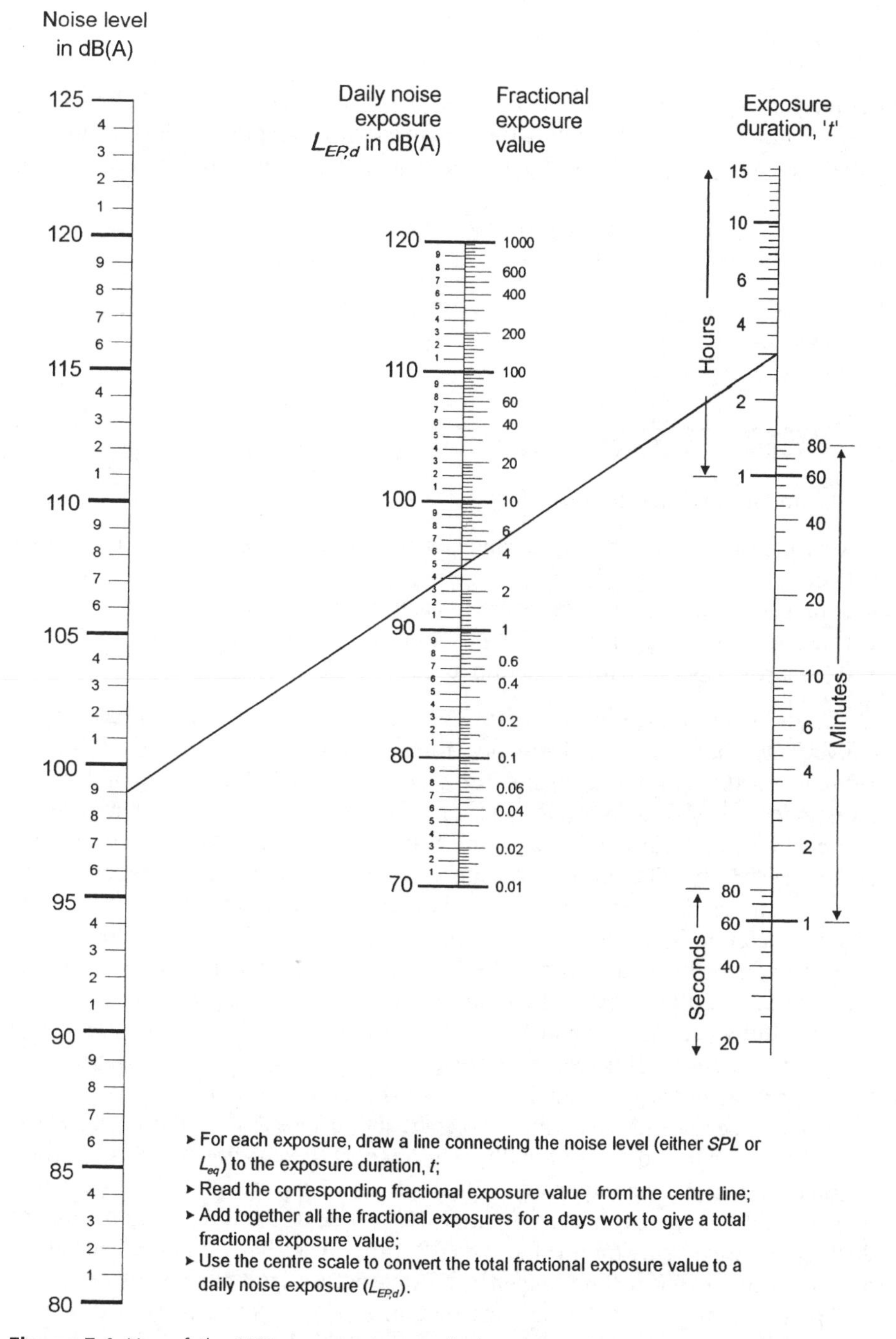

Figure 5.1 Use of the HSE nomogram to calculate $L_{EP,d}$. From L108 Reducing Noise at Work (HSE, 1998). © Crown copyright material is reproduced with the permission of the Controller of HMSO and Queen's Printer for Scotland.

Example

A machine operator is exposed to an $L_{AEP,d}$ of 83 dB 4 days a week. On Fridays, significant time is spent cleaning the machines with high pressure hoses and on that day the $L_{AEP,d}$ is 89 dB. What is the average $L_{AEP,d}$ over the week?

$$L_{EP,d} = 10 \times \log\left(\frac{\left(4 \times 10^{\frac{83}{10}} + 10^{\frac{89}{10}}\right)}{5}\right) = 85.0 \approx 85\,\text{dB}$$

Percentage dose

The simplest approach to noise dose assessment is to:

- Measure the A weighted L_{eq} to which an employee is exposed for each part of a shift;
- Establish the duration of exposure to each of these levels;
- Use the L_{eq} measurements and exposure times to calculate the overall $L_{AEP,d}$;
- Compare this value with the action levels.

Other approaches have been used in the past and it is sometimes necessary to convert measurement data from one form to another. Up to the 1980s, the recommended maximum exposure in the UK was the same as the second action level under the Noise at Work Regulations. This level of exposure was taken to be 100 per cent, and daily noise exposure was commonly expressed in terms of a percentage dose. The output of most personal dosemeters was expressed as a percentage of this exposure. One advantage of this system was that percentage doses can easily be added. An employee receiving a 75 per cent dose in the morning and a 33 per cent dose in the afternoon will have a daily dose of 108 per cent. Modern equipment can often output data as a percentage as well as in other formats, but since the Noise at Work Regulations have two action levels, some ambiguity can result. If percentages are used, it is essential to specify whether this is a percentage of the first action level or of the second action level. For consistency, it has been normal to define the second action level as representing a 100 per cent dose. With the advent of the Physical Agents (Noise) Directive this situation becomes much more complicated. There are now two additional $L_{EP,d}$ values which could be regarded as representing 100 per cent; the lower action value of 80 dB and the exposure limit value of 87 dB. Moreover, the original 100 per cent – an $L_{AEP,d}$ of 90 dB – will no longer be of any special significance.

The equations given below assume that an $L_{AEP,d}$ of 90 dB is equivalent to a percentage dose of 100 per cent. They can easily be amended for use with alternative conventions.

To convert an $L_{EP,d}$ to a percentage dose (based on 8 h at 90 dB = 100 per cent):

$$\text{Percentage dose} = 100 \times 10^{\left(\frac{L_{EP,d}-90}{10}\right)}\% \qquad (5.7)$$

To convert a percentage dose (based on 8 h at 90 dB = 100 per cent) to $L_{EP,d}$

$$L_{EP,d} = 90 + 10 \times \log\left(\frac{\%\ \text{dose}}{100}\right) \text{dB} \qquad (5.8)$$

Examples

1. What percentage dose (based on the second action level being equal to 100 per cent) corresponds to an $L_{AEP,d}$ of 94.5 dB?

$$\text{Percentage dose} = 100 \times 10^{\left(\frac{L_{EP,d}-90}{10}\right)} = 100 \times 10^{\left(\frac{94.5-90}{10}\right)}$$
$$= 100 \times 10^{0.45} = 282\%$$

2. What percentage dose would the same personal daily exposure correspond to if 100 per cent is assumed to be equivalent to a personal daily exposure of 87 dB%?

$$\text{Percentage dose} = 100 \times 10^{\left(\frac{L_{EP,d}-87}{10}\right)} = 100 \times 10^{\left(\frac{94.5-87}{10}\right)}$$
$$= 100 \times 10^{0.75} = 562\%$$

Here, 87 has replaced 90 in equation (5.7).

3. What $L_{AEP,d}$ corresponds to a percentage dose of 132 per cent? (Assume an $L_{AEP,d}$ of 90 dB is 100 per cent)

$$L_{EP,d} = 90 + 10 \times \log\left(\frac{132}{100}\right) = 90 + 10 \times \log 1.32 = 91.2\,\text{dB}$$

Sound exposure, E_A

The sound exposure, which is measured in units of pascal-squared hours, has never really caught on. The scale has two advantages over others:

- It is a linear scale, so 0.5 Pa^2-h represents half the dose represented by 1.0 Pa^2-h. Doses in Pa^2-h can be added arithmetically.
- 1.0 Pa^2-h is the same as an $L_{EP,d}$ of 85 dB, the first action level in the 1989 Noise at Work Regulations.

To convert $L_{EP,d}$ into an exposure in Pa^2-h:

$$E_A = p_0^2 \times 8 \times 10^{\frac{L_{EP,d}}{10}} \qquad (5.9)$$

where E_A is the sound exposure, measured in pascals squared-hours, and

p_0 is 2×10^{-5} Pa

Conversely,

$$L_{EP,d} = 10 \times \log\left(\frac{E_A}{8 \times p_0^2}\right) \qquad (5.10)$$

Examples

1. What sound exposure corresponds to an $L_{AEP,d}$ of 86 dB?

$$E_A = p_0^2 \times 8 \times 10^{\frac{L_{EP,d}}{10}} = \left(2 \times 10^{-5}\right)^2 \times 8 \times 10^{\frac{86}{10}} = 1.27\,\text{Pa}^2\text{-h}$$

2. If the sound exposure of an employee has been assessed at 1.6 Pa^2-h, what is this person's $L_{AEP,d}$?

$$L_{EP,d} = 10 \times \log\left(\frac{E_A}{8 \times p_0^2}\right) = 10 \times \log\left(\frac{1.6}{8 \times (2 \times 10^{-5})^2}\right) = 87.0\,\text{dB}$$

Ranking different exposure periods

It can sometimes be useful if an employee moves between a number of different noise environments in the course of a working day to assess the relative contribution of each task to the daily noise exposure. There are a number of approaches to this, but the simplest (because it does not introduce any new calculation techniques) is to consider each task in turn. Work out what the $L_{EP,d}$ would be on the assumption that this task represents the only significant noise exposure during the shift. The set of figures that result are sometimes called partial $L_{EP,d}$s, and their relative magnitudes reveal the relative contributions to the daily noise exposure. Clearly, any attempts to reduce the daily noise exposure will most usefully be concentrated on those parts of the day when exposure is greatest.

Example

An employee's daily noise exposure is tabulated below. (a) Work out the corresponding daily exposure, $L_{EP,d}$. (b) List the machines in the order of their contribution to the daily exposure.

Machine	Level	Time
Thicknesser	102 dB	20 min
Frame saw	96 dB	1 h
Panel saw	94 dB	3 h
Router	94 dB	2 h

(a)

$$L_{EP,d} = 10 \times \log\left(\frac{\left(t_1 \times 10^{\frac{L_1}{10}} + t_2 \times 10^{\frac{L_2}{10}} + t_3 \times 10^{\frac{L_3}{10}} + t_4 \times 10^{\frac{L_4}{10}}\right)}{8}\right)$$

$$= 10 \times \log\left(\frac{\left(1/3 \times 10^{\frac{102}{10}} + 1 \times 10^{\frac{96}{10}} + 3 \times 10^{\frac{94}{10}} + 2 \times 10^{\frac{94}{10}}\right)}{8}\right)$$

$$= 94.4\,\text{dB}$$

(b)

Thicknesser: $L_{EP,d} = L_{eq} + 10 \times \log(\frac{t_m}{8}) = 102 + 10 \times \log(\frac{1/3}{8}) = 88.2\,\text{dB}$

Frame saw: $L_{EP,d} = L_{eq} + 10 \times \log(\frac{t_m}{8}) = 96 + 10 \times \log(\frac{1}{8}) = 87.0\,\text{dB}$

Panel saw: $L_{EP,d} = L_{eq} + 10 \times \log(\frac{t_m}{8}) = 94 + 10 \times \log(\frac{3}{8}) = 89.7\,\text{dB}$

Router: $L_{EP,d} = L_{eq} + 10 \times \log(\frac{t_m}{8}) = 94 + 10 \times \log(\frac{2}{8}) = 88.0\,\text{dB}$

So the greatest contribution comes from the panel saw, followed by the thicknesser, the router and the frame saw.

A more traditional approach to the above type of problem would be to work out a percentage dose for each operation.

Time limits

Sometimes the level to which an employee is to be exposed is known, and the problem is to decide on a daily time limit which ensures that daily exposure will be below one of the action levels.

$$L_{EP,d} = L_{eq} + 10 \times \log\left(\frac{t}{8}\right) \tag{5.11}$$

Here, t is the time limit to be determined. The $L_{EP,d}$ to be used is the action level below which it is required to keep exposure, while the L_{eq} is the level to which the operator of a particular machine is exposed. Numbers can be inserted for L_{eq} and $L_{EP,d}$, following which the equation can be solved for t. Or the equation can first be re-arranged into the form

$$t = 8 \times 10^{\frac{L_{EP,d} - L_{eq}}{10}} \tag{5.12}$$

Examples

1. A night club employee is exposed to an L_{eq} of 102 dB during a 3 h shift. What is this employee's personal daily dose, $L_{AEP,d}$?

$$L_{EP,d} = L_{eq} + 10 \times \log\left(\frac{t}{8}\right) = 102 + 10 \times \log\left(\frac{3}{8}\right) = 97.7\,\text{dB}$$

2. For how long can the employee above be employed before the second action level is reached?

$$t = 8 \times 10^{\frac{L_{EP,d} - L_{eq}}{10}} = 8 \times 10^{\frac{90-102}{10}} = 8 \times 10^{\frac{-12}{10}} = 0.5\,\text{h}$$

Residual time limits

Occasionally, a certain amount of noise exposure has already occurred and it is necessary to decide on a time limit for a further period of exposure to a different level. There are a number of approaches to this calculation, none of them particularly simple. One approach is to use the equation for calculating $L_{EP,d}$ from two different levels of exposure.

$$L_{EP,d} = 10 \times \log\left(\frac{\left(t_1 \times 10^{\frac{SPL_1}{10}} + t_2 \times 10^{\frac{SPL_2}{10}} + \ldots.\right)}{8}\right) \tag{5.13}$$

In this case, t_1 and SPL_1 are known values which represent the previous exposure. SPL_2 is the future exposure level, $L_{EP,d}$ is the action level below which the daily exposure is required to remain. If these values are inserted, then the equation can be manipulated to yield a value for t_2, the permitted time limit for the remaining exposure.

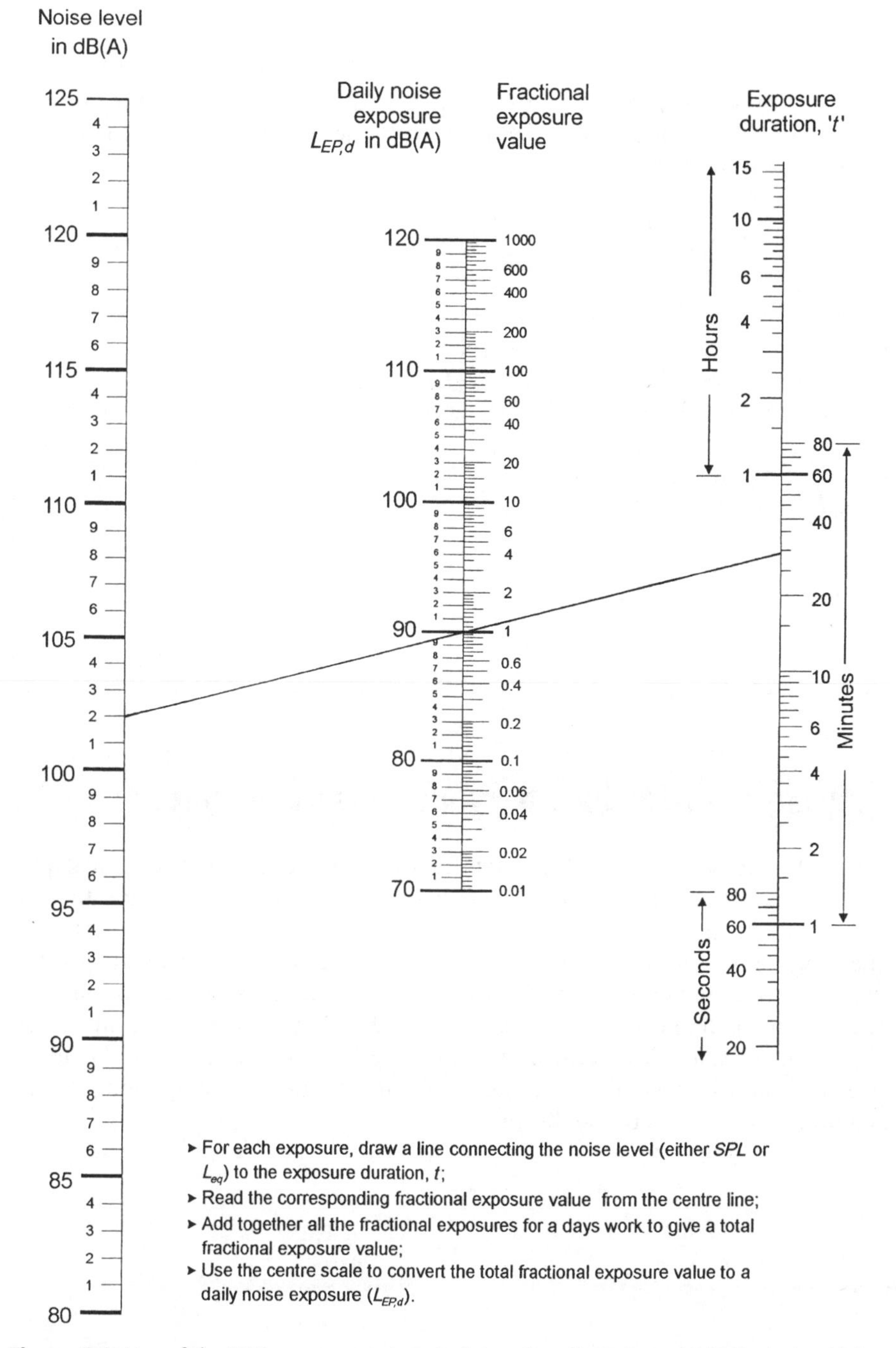

Figure 5.2 Use of the HSE nomogram to calculate a time limit. From L108 Reducing Noise at Work (HSE, 1998). © Crown copyright material is reproduced with the permission of the Controller of HMSO and Queen's Printer for Scotland.

Example

An employee has spent 3 h exposed to a level of 92 dB. For how long during the same shift can this employee be exposed to a level of 89 dB while remaining below the second action level?

$$L_{EP,d} = 10 \times \log\left(\frac{\left(t_1 \times 10^{\frac{L_1}{10}} + t_2 \times 10^{\frac{L_2}{10}}\right)}{8}\right)$$

$$90 = 10 \times \log\left(\frac{\left(3 \times 10^{\frac{92}{10}} + t_2 \times 10^{\frac{89}{10}}\right)}{8}\right)$$

$$10^9 = \frac{3 \times 10^{9.2} + t_2 \times 10^{8.9}}{8}$$

$$t_2 = \frac{8 \times 10^9 - 3 \times 10^{9.2}}{10^{8.8}} = 5.1\ \text{h}$$

Using the HSE chart to work out a time limit

When it is necessary to work out the limit on time to be spent in a particular environment, it can be considerably easier to use the HSE nomogram than to carry out the same calculation on a calculator. Figure 5.2 demonstrates this for the simple situation used in a previous example where a night club employee is exposed to a sound pressure level of 102 dB. This sound pressure level is found on the left hand column, while the target $L_{EP,d}$ – 90 dB in this case – is identified on the left hand side of the central column. The line joining these two points is extended until it meets the right hand column at a point corresponding to the required time limit – around 30 min.

Assessment of noise exposure

Is a survey necessary?

When planning a noise exposure survey, it is first of all necessary to be very clear about its scope. It may be required to assess the noise exposure of all employees in an establishment, or of those working in one or more departments or trades within it. It may be that the noise exposure of a number of individuals is to be assessed, or the exposure of a group carrying out a similar range of tasks.

The Noise at Work Regulations require an employer to arrange a noise exposure survey when it appears likely that any employee is exposed above the first action level, or above the peak action level (it is technically possible for the peak action level to be exceeded even if the first action level is not), and the starting point will normally be when there is some evidence that this level of exposure is taking place. This evidence might be:

- Evidence that one or more employees is exposed to an A weighted sound pressure level above 85 dB for a significant period of time. This might follow from some initial measurements, or it may be deduced from a simpler test such as that recommended by the Health and Safety Executive (1998); if voices need to be raised when communicating over a distance of 2 m, then the sound pressure level is likely to be greater than 85 dB.
- Health surveillance results from one department or trade suggesting progressive hearing loss.
- The industry concerned, or the machinery in use, or the particular tasks carried out, may be ones which are known to frequently involve noise exposures in excess of one of the action levels.

On the other hand, there will normally be employees whose noise exposure is clearly well below any action level. The noise exposure of the majority of office staff, for example, would not be expected to approach either the first or the peak action level. Noise levels here may well be considered as part of a routine risk assessment, and under the Physical Agents (Noise) Directive it is a duty of employers to assess the noise exposure of all employees. However, this assessment would not normally involve any physical measurements.

The Noise at Work Regulations place the onus on the employer to find a competent person to carry out the assessment. The concept of a competent person is a recurring one in health and safety management. Competence is established by a combination of formal training, qualifications and practical experience. In the case of workplace noise assessments it is very likely that a competent person will have successfully completed a course in workplace noise assessment, but no such qualification on its own makes its holders 'competent', and a skilled, experienced person may be competent to carry out noise assessments without any formal qualification. An individual may be perfectly competent to carry out the majority of workplace noise assessments, but there are a number of situations requiring special skills and experience. A competent person is expected to know how far his or her competence extends and be able to advise when more specialized skills are needed.

Planning the survey

Some of those who carry out noise assessments work for the same organization, while others come from outside as consultants. In many cases the situation will fall between these extremes as different departments or divisions recharge for their work. Whatever the precise arrangements, it is desirable for the assessment to be carried out as speedily and efficiently as is consistent with making a satisfactory assessment of the exposure of everyone covered. Although some idea of what is involved can be gathered in telephone conversations, it will normally be necessary to visit an unfamiliar workplace first to collect information about the scope of the assessment. Only then will it be possible to estimate the time required and – for external assessors – the cost involved.

The sort of information to be collected on the initial visit would include the numbers of employees to be included, the shift patterns, whether or not the work carried out on different shifts was the same (i.e. is a separate assessment required for each shift, or can data on shift hours be combined with measurements made on the most convenient shift?). Any particularly difficult measurement situations can be identified, and it will become clear if groups of employees have essentially the same job and can therefore be assessed as a group. This is a good time to ask for a copy of any previous noise assessments. If they are accurate they will help to concentrate efforts on those areas where noise exposure is near or above the action levels. If a set of plans of the workplace are available they may save considerable time in recording measurement positions. In some cases the 'workplace' is a fluid concept. Measurements may need to be made over a considerable geographical area.

Some jobs vary a great deal from one day to the next, and it may be necessary to sample noise levels on more than one day.

Following the initial visit, it should be possible to plan how the data collection will be carried out, and it is only at this stage that it will become clear how much time needs to be allocated to carry out the survey. Two sorts of data will need to be collected. Sound pressure levels measured at appropriate positions, and also the kind of information that can be acquired by interviewing workers, supervisors and health and safety managers – information such as the hours of work, and whether training has been given in the effects of noise, variations in the work patterns and in the types of machinery in use, and details of any hearing protection which is provided.

The measuring equipment to be used can be decided. In some cases – in hazardous areas, for example – it may be necessary to hire equipment specially for the job.

The relative merits of hand-held sound level meters and personal dosemeters are a matter of debate. It is possible to get incorrect results with dosemeters for several reasons:

1. The practical problems of attaching the microphone mean that it is normally mounted rather closer to the wearer's body and/or clothing than is ideal from a measurement point of view. Reflections from and/or screening by the wearer's body can lead to significant errors.
2. The microphone may brush against clothing, and as a result record higher sound levels than the wearer is actually exposed to.
3. Because the measurement is not directly supervised by the assessor, the measurement may be interfered with. Although there are normally ways of preventing direct tampering, it would not always be apparent if the dosemeter had been taken off and left in a noisy (or quiet) place, or exposed to unrepresentative noise levels in situ.

On the other hand, there are situations where it is not very practical for an assessor to stand close to an employee. The noise dose of a fork-lift truck driver, for example, would be difficult to measure safely with hand-held equipment. It is also useful that several employees can wear dosemeters which can all be fitted by a single assessor. The type of dosemeter which at the same time as measuring the noise dose also records a time history of noise exposure for download to, and analysis on, a computer can yield a great deal of information about noise exposure patterns which may not otherwise be available. It can also be useful in identifying, and removing from the measurement, periods of time when noise levels were deliberately falsified.

Some of the disadvantages of dosemeters can be avoided by using the type of instrument which can be made to record a time history of the complete measurement. This will normally need to be downloaded to a computer for analysis. It is then, for example, possible to recalculate the L_{eq} after excluding a period when the dosemeter was not being used correctly. Alternatively, the time history may identify periods when the sound levels were higher or lower than had been

Table 6.1 Sound level meters and dosemeters compared

Sound level meter measurements	Dosemeter measurements
Only one measurement at a time	One operator can set several dosemeters going
The exposure of a number of operators can be sampled during a single shift	Data only collected on the noise exposure of the wearer
Extraneous noise easily identified and excluded from the measurements	Normally no record of unwanted events included in the measurement
Under the direct control of the operator	Wearer may try to tamper with data
Possible risks to person carrying out measurements	An uncontrolled cable may cause a hazard to the wearer
Predictions can be made about the effect of possible variations in the work pattern	Only data relating to the actual work done are recorded
The main noise sources will normally be obvious	No information about the source of the noise exposure
Capable of more accurate sound level measurements	Accuracy limited by proximity to the wearer's body

expected. In this case further measurements can be made. The ideal would be to use a combination of dosemeter and hand-held measurements. When analysing the results it should be possible to show that the results are consistent. If they are not consistent, then the reason may be relevant. It may be because the dosemeter was not worn for the whole shift, but it may be because the wearer was exposed to a noise source of which the assessor was previously unaware.

It is good practice to prepare a set of printed survey sheets. These will be kept for many years in case the original measurement data need to be consulted for any reason. As much information as possible can be printed on each sheet to reduce the amount of writing to be done on site under possibly difficult conditions – poor light, falling rain, etc. For example, before starting work on the project the following could be printed on each sheet:

Workplace information (company, site, etc.)

- Date
- Assessor identification
- Equipment used with serial numbers and calibration details
- Questions to be asked

Boxes can be provided for

- Employee name
- Machine description and identification
- Material being worked
- Measurement time
- A weighted L_{eq} and linear peak levels
- Answers to standard questions to be asked.

This approach has the advantage that it reminds the assessor to collect all the information required. It also means that the same information appears in the same place for each person assessed, so that comparing and collating results afterwards is much easier. For this reason, it is useful if space is provided for the outcome of the assessment on that particular employee; in other words the values of the assessed $L_{EP,d}$ and L_{Cpeak} values and the action levels exceeded.

Noise assessment

#

Date		Time	
Department/section			
Measuring equipment			

Tool type	
Make/model	
Serial #	
Process/operator	
Material/workpiece	

Noise measurements

Run	Time	*L/R*	L_{Aeq}	L_{Cpk}	Comments

Plan

Hours		Breaks		O/T?
Time/day	Typical		Maximum	
Time/operation				
Operations/day	Typical		Maximum	
Other work				
HP type			Condition	
Training/info				
Noise problem?				

Backstone Sawmills noise assessment *April 200X*

Figure 6.1 A noise survey sheet.

Space should be left for a rough plan of the surroundings – showing the position and approximate distance of the workstation and the main noise sources – and for notes about special circumstances which are relevant to the assessment and which were not foreseen.

Figure 6.1 shows an example of a noise survey sheet. It is not definitive; it may work well in some workplaces but be totally inappropriate in others.

Measuring sound pressure levels

The standard procedure for measuring the sound pressure levels to which an employee is exposed is to measure at the position of the centre of the employee's head in the absence of the employee, but under normal noise conditions. This convention arises from the original research to establish the relationship between noise exposure and hearing loss, possibly due to the limitations of the equipment in use at that time. It is very often impractical to make this type of measurement simply because in the absence of the employee, noise from the employee's own work will either not be typical or indeed will not be present at all. Instead, it is more normal to measure close to the employee's ear while he or she continues to work normally. This will yield results which under normal conditions will be very similar to those that would be measured as described above. Some exceptions to this are discussed later in this chapter.

In most cases the exposure of each ear must be measured, since there may be a considerable difference in cases where the main noise source is to one side. The highest of these two levels will then be used in the assessment. It is not normally practical to measure an employee's exposure during a complete shift. If the noise level varies little during the shift it is necessary to measure for long enough to acquire a representative sample of the noise levels. If there is a cyclical variation in sound pressure level, it is normally necessary to measure over a number of complete cycles. For example, a task might involve picking an object from a pallet, clamping it in place, machining it, and then removing and stacking the finished piece. The whole process might take a few seconds or several minutes and the minimum measurement period would be for one complete cycle. Unless the time taken were very long, it would be more satisfactory to measure over at least two full cycles.

Where the job involves spending different periods on completely different tasks, then it will be necessary to measure levels during each different task and later use these along with information about the time spent on each to calculate the overall $L_{AEP,d}$.

Wherever possible, measurements should be of A weighted L_{eq} and of C weighted peak level (if C weighting is not available then Z or linear weighting may be acceptable alternatives). Many modern sound level meters will measure these quantities simultaneously (dosemeters nearly always do), but if this is not possible then they must be measured separately.

Most noise exposure assessments will nowadays be made with an integrating sound level meter. If one is not available, then it is just about possible to measure sound pressure levels satisfactorily with a nonintegrating sound level meter as long as these levels do not fluctuate greatly.

If the effectiveness of the hearing protection in use is to be assessed, it will be necessary to measure the C weighted L_{eq} and possibly also the octave band sound pressure levels. This quantity of data may be more easily stored in a sound level meter's memory for later retrieval. The relevant data file number will then of course have to be recorded along with other data about the measurement position, time and noise sources.

Although most measurements will be made as employees go about their normal duties, it may be necessary for machines to be started up specially so a complete set of measurements is obtained. In this case they should be used during the measurements to work the same types of material as they are used for in practice. An idling machine will normally be much quieter than when it is being used normally.

Collecting information

Information about exposure times, working practices and use of hearing protection can be obtained in a number of ways:

- Interview the staff concerned
- Interview managers and/or supervisors
- Direct observation
- Consult written records.

Direct observation normally leads to reliable information except in so far as it may only cover a short and possibly unrepresentative period. Written records (of output, machine operating times, etc.) will probably be available in a factory environment, and can then be useful for confirming exposure times. Written records of other noisy work – in many outdoor occupations, for example – are much less likely to exist. Information given by employees or by managers may intentionally or unintentionally be biased. The best option is to collect evidence in as many ways as possible. Hopefully the various sources will agree, but sometimes it is necessary to resolve contradictions by making further enquiries.

Information about the availability and use of hearing protection can be observed directly and compared with the employer's stated policies.

Assessing noise doses

By the time work on site finishes, information will have been collected for each of the employees covered about noise levels, the time for which the employee is exposed to each level, variations in shift patterns and noise exposure, and other relevant details such as the use of hearing protection. Whether an employee is assessed as an individual or included in a group for whom a collective assessment is made, it should now be possible to calculate the employee's personal daily exposure, $L_{EP,d}$ (see Chapter 5). In many cases it will be necessary to calculate a number of different values of $L_{EP,d}$ to take account of variations in shift patterns, overtime, product mix and allocation to different duties. In each case a separate assessment will be made of peak exposure for comparison with the peak action level.

Reporting on an assessment

Many organizations have a house style for reports, and there is certainly no single format for noise at work assessments. It may be appropriate to use a different format depending on the scale of the assessment and the nature of the findings. Alternatively, the employer involved or the organization responsible for the noise exposure assessment may have particular requirements. It should be borne in mind that the report submitted will be kept by the employer until a new assessment has been completed. It may be shown to enforcement officers, and it may even resurface as evidence in a compensation case many years hence. There are certain features which are essential to any report, whether it is produced within an organization or by an external consultant:

1. It must clearly identify the scope of the assessment in terms of the sites, departments and individuals covered.
2. The measured levels and the exposure times estimated must be stated, although for a large or complicated assessment this detailed information is probably best put in an appendix.
3. For each employee an $L_{EP,d}$ – or a range of $L_{EP,d}$s to cover different circumstances – must be stated.
4. These $L_{EP,d}$s should be explicitly interpreted so it is clear which action level(s) are exceeded by which employees.
5. A similar interpretation must be made of peak exposures.
6. The duties of the employer with regard to each employee should be pointed out.

Normally some suggestions will be included as to how the employer could comply with the legal duties. These may include engineering measures such as enclosure of noisy machines, management measures such as changing the allocation of employees to different duties, and longer term measures such as investing in new machinery. The competent person should aim to point out a range of possibilities rather than telling managers to go down one particular road. The duty rests at the end of the day on the employer who will weigh up benefits against costs, impact on the work process, etc., before deciding which measures to adopt.

It is highly likely that recommendations will be made about the provision and management of hearing protection. Although personal protective equipment (PPE) such as this is widely described as 'the last resort', chronologically it is the first action which is normally taken. Hearing protection can be provided within hours of it becoming clear that there is a problem, while most other actions to reduce noise exposure will take weeks or months to implement. The survey sheets used during data collection should not be included in the report. They must be kept as a permanent record by the individual or organization responsible for carrying out the assessment. Most employers would prefer a clear summary of the findings to a mass of measurement data, and those who wish to see the original survey sheets can be supplied with a copy.

Some difficult measurement situations

Some noise assessments have unique features which may challenge the ingenuity of a competent noise assessor. Work in confined spaces, operation of vehicles and work with animals are examples of situations which require an extension of the ordinary techniques. There are a number of work situations in which it is particularly difficult to make useful measurements of the noise levels to which an individual is exposed. Measurement in such cases is a highly specialized matter which will in most cases be referred to a specialist. If the dominant noise source is thought to be close to the employee's ear (i.e. within a few centimetres) then the normal technique of measurement close to the head will not yield satisfactory results. This is the normal situation when the person concerned is wearing headphones, for instance aircraft crews and the growing number of call centre workers. Two techniques may help here: one is to use a head and torso simulator or HATS (Figure 6.2). This is a dummy head and shoulders to which the headset can be fitted. It then needs to be fed with the same signal as would be heard by the real employee. The other approach is to use a miniature microphone inside the

Figure 6.2 A head and torso simulator (HATS).

ear canal. Both require a great deal of knowledge and experience. Fortunately, a number of specialist suppliers of equipment for use in call centres have invested considerable time and money in developing workstations whose output levels are controlled so as to fall below those which would be likely to pose a threat to the hearing of the operators. This is a situation where satisfactory noise exposure assessments can be carried out mainly by using data supplied by the manufacturers rather than by making direct measurements.

Motorcyclists are also exposed to noise which is generated close to their ear, and the measuring difficulties here are compounded by various safety considerations.

When the sound energy is transmitted mainly by bone conduction, as is the case, for example, with divers, then the normal measurement procedures will again not be applicable. Here, too, exposure assessments are likely to be based mainly on published information.

Case study 6.1 A workplace noise assessment

Backstone Sawmills Ltd manufactures components for the furniture industry. Around 40 shop-floor employees work in a large, modern building operating machines which convert rough-sawn timber into finished components, normally going through three or four stages on the way. A separate packing and despatch building carries out a number of finishing processes, as well as loading the components on to pallets for transport to the customer. Three forklift truck drivers work both outside and inside, and two engineers are responsible for maintenance work throughout the factory.

Preliminary survey

During a preliminary survey, it was established that large batches of components are produced, so that each employee is likely to be assigned to one particular machine for a large part of the shift. The main factory is large, and the building surfaces tend to reflect sound so that a significant reverberant sound field is present, affecting even those employees not directly working at a noisy machine. Workers work 8-h shifts from Monday to Thursday, but finish at 1 p.m. on Fridays. Overtime is sometimes worked on Friday afternoons and Saturday mornings. However, it was clear that Friday would be a bad day to carry out a noise exposure survey.

Planning

It was decided that exposure of the employees in the main factory could be carried out by measuring sound pressure levels at each identified workstation, and by collecting information from the operators about working practices. Forklift drivers spend a significant proportion of their day working in the factory, and even though the yard area is much quieter, it was decided to ask two drivers to wear personal dosemeters to assess their noise exposure. The working patterns of the engineers was likely to be more complicated, and time was allocated at this stage to collect information about their typical working patterns.

In the packing building, sound pressure levels seemed to be lower, but enough measurements would need to be made to confirm or disprove this. Two days were allocated to carry out the complete survey.

First day

On the first day, two forklift drivers agreed to wear dosemeters. These were fitted to them an hour after the start of the shift. The purpose of the instrument was explained, as were the precautions they were asked to take to ensure that a correct reading was obtained. They were further requested to keep the dosemeter on until it was collected from them in the middle of the afternoon. Measurements were made at as many working positions as possible. This included a number of representative positions away from machinery so that an estimate could be made of the noise exposure of the labourers whose job is to sweep up and keep the shop floor tidy. In each case, information was collected from the operators about the times they spent operating that machine, details of any other machines they operated, and how long they might spend on each machine in the course of a single shift. They were asked if they used hearing protection, and observations were made about whether they were in fact doing so, whether it was correctly worn, along with the arrangements for providing it and, in the case of ear muffs, their condition. The dosemeters were retrieved from the forklift drivers before the end of the shift. In one case, the reading was used to calculate the personal daily dose over a full shift, and this was consistent with the proportion of time observed to be spent working inside the factory. The second dosemeter showed a surprisingly high reading both for L_{Aeq} and for L_{Cpeak}. The wearer agreed that he 'might have accidentally knocked the microphone'. At the end of the shift, the information collected was collated and priorities were established for the second day of measurements.

Second day

On the second day, dosemeters were fitted to the driver who had previously produced the unexpected readings, and also to the third, previously unmonitored, driver. The remainder of the working positions in the main factory were surveyed, and the rest of the morning was spent with the engineers, establishing typical patterns for their noise exposure. Initially, they argued that every day is different. After discussing the work carried out over the previous week, it became clear that certain patterns could be established. Noise levels were measured in the engineers' workshop while they sharpened and repaired tools. They discussed which parts of the factory claimed most of their attention, and they pointed out if they spent a long period in one part of the factory it was normally because a particular machine needed extensive maintenance work and it would normally not be operating at the time (although nearby machines probably would be in use). A set of work patterns representing typical days was developed and the likely personal daily dose according to each of these scenarios was calculated.

At the end of the morning, the dosemeters were collected from the forklift drivers, and it was discovered that this time the calculated daily doses were consistent with each other and with the first of the doses measured on the forklift drivers the previous day. The factory lunch break was spent collating the data collected so far. Repeat visits were then made to some workstations where the available information seemed to be incomplete or contradictory. Most of the rest of the shift was spent in the packing and despatch building. It was discovered that the most important noise sources were nail guns used to assemble pallets and packing cases. When these were being operated, sound pressure levels were between 93 and 95 dB. Otherwise they were generally below 80 dB. It was therefore important to establish the periods for which the nail guns were used, and who was likely to be nearby at these times. This information was first sought from the operators, and despatch records were then consulted – along with some measurements of the time taken to assemble each pallet – to confirm both the average and the greatest periods spent using them in the course of a day. The peak levels measured when these nail guns were in use, between 120 and 125 dB, were the highest found in the factory as a whole. It was noted, though, that they were comfortably below the peak action level under the Noise at Work Regulations 1989 and also the lower peak exposure action value of the Physical Agents (Noise) Directive.

It was noticed that the quiet end of this building was – apparently for historical reasons – well supplied with hearing protection signs which were otherwise absent from the premises.

Post-measurement analysis

At the end of the shift, the factory manager was given a preliminary verbal report of the findings, although it was emphasized that any information given at this stage was subject to confirmation.

After calculation of the personal daily exposure of each employee, it was found that at the majority of the working positions in the main factory an exposure above the second action level was likely. At a few positions the exposure was likely to fall between the first and second action level. It was recommended that the entire building should be made a hearing protection zone. In the packing/despatch building, it was established that the two nail gun operators would, on some days, be exposed above the second action level. It was decided that hearing protection would be compulsory when using these machines, and also for any others who at the time were within an area to be clearly demarcated.

If the engineers observed these hearing protection zones, then they would not be exposed above the first or second action level as a result of any reasonably foreseeable work pattern. Forklift drivers would similarly be required to use hearing protection inside the main factory, although concerns were raised about the practicality and safety of this.

Training sessions on the effects of noise and ways of preventing hearing damage were required. These would be supplemented by notices identifying the new hearing protection zones, and by other information about protecting hearing displayed on notice boards. Suggestions were made about ways to improve arrangements for obtaining and storing hearing protection.

The changes in action levels under the forthcoming Physical Agents (Noise) Directive were pointed out to the factory management, and it was recommended that over the intervening period various steps were considered to reduce noise levels in the factory to comply with existing duties, and to avoid the need for more stringent hearing protection requirements when the new directive comes into force.

Hand–arm Vibration

Fundamentals of vibration

What is vibration?

Vibration occurs when a body oscillates repeatedly around an equilibrium position. The simplest form of vibration can be observed when a small mass is hung from a spring. Left to itself, the mass will come to rest at the position where the gravitational force acting on it is equal in size (but opposite in direction) to the force exerted by the stretched spring (Figure 7.1).

Displace the mass slightly from that position – either up or down – and the forces are no longer in balance. For example, if it is pushed slightly downwards and then released, then the spring is stretched more than before, and the force it exerts is then greater than the gravitational force. A net upwards force exists and the mass will accelerate upwards. Eventually it reaches the equilibrium position again, but it is now moving and it is only when it has passed the equilibrium position that a net downwards force will exist which can start to slow the mass down. Eventually it comes to a halt. It is now above its equilibrium position, and the gravitational force exceeds that exerted by the spring. The mass will be accelerated downwards. It will continue to move upwards and downwards about the equilibrium position until an external force acts to change this motion.

This sort of motion is known as simple harmonic motion (SHM for short). It is a very simple model of what happens in real systems. Even a simple system of a real mass and real spring does not always behave exactly like this. For example, the mass may start to rotate slightly as well as moving up and down. Nevertheless, it can be applied to a great number of real systems in order to begin to understand what is happening when a real object vibrates.

The vibration will have a frequency. This is the number of complete oscillations that occur in one second and, as with sound, the unit it is measured in is the hertz. The mass and spring are likely to have rather a low vibration frequency – probably no more than 1 Hz. Most vibrating bodies will vibrate at higher frequencies than

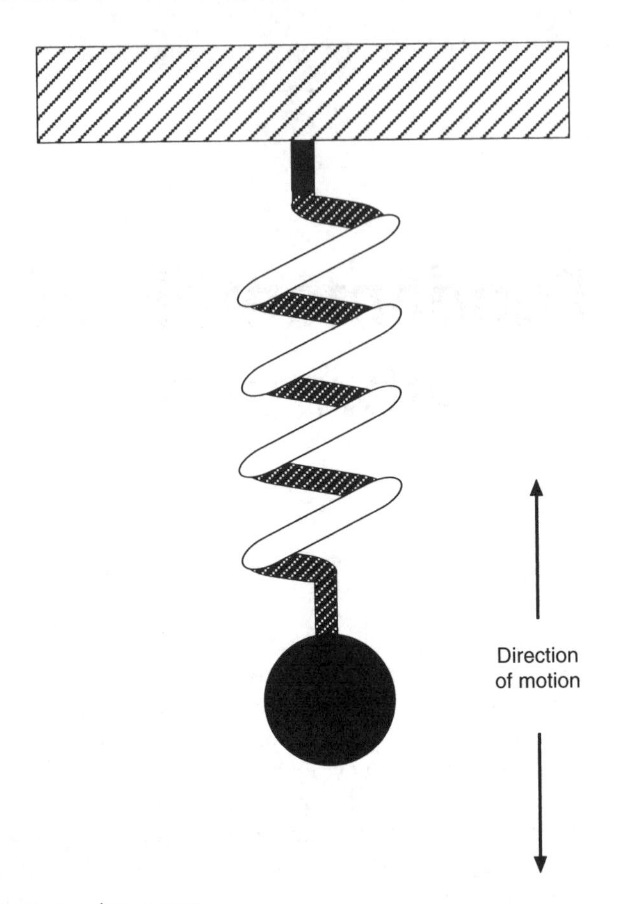

Figure 7.1 A mass-spring system.

this, maybe up to hundreds or even thousands of hertz. Overall, though, vibration affecting human beings tends to have a frequency range which is rather lower than the range of frequencies concerned with audible noise. They overlap, though vibrations of interest for their health effects commonly have a frequency range from 1 Hz up to 1000 Hz.

It is tempting to see vibration as essentially static, while noise is a manifestation of a travelling wave. This is too simplistic as vibrations can certainly travel. If this was not the case then a vibrating mechanism inside a tool would have no way of causing the tool handle to vibrate and vibration energy would not be transferred to the human body. With sound, standing waves can be established in an enclosed space, so that the theoretical differences between sound and vibration are not particularly great. In practice, though, waves travel much faster in solids than in the air. The motion of vibration in workplace situations is not normally apparent, and it is reasonable to treat vibration for most of the time as though it were a static phenomenon.

Damping and isolation

It was stated above that the mass and spring system would carry on oscillating until an external force acted to change its motion. In practice, external forces are always present which act to oppose the motion of a vibrating body. They are known as damping forces. In this case, the resistance of the air to the motion of the mass is probably the most important source of damping. There are others; the spring will not be perfect and a small amount of energy will be converted to heat every time the spring stretches and contracts. If the point of support is not absolutely stationary, then there will also be a transfer of energy to the support system. Whenever energy is removed from a vibrating system, damping occurs and tends to reduce the amplitude of the vibrations. The vibration of a hand-held tool will be damped by various internal processes, by contact with the work, and by transfer of energy to the hand of the operator, as well as by any air damping that occurs. Forced vibrations occur when the energy removed by damping from the vibrating system is replaced from an energy source. An equilibrium is established whereby the energy supplied from the source is equal to the energy removed by damping. Thus the vibration at the handle of a tool operated under uniform conditions will tend to be steady over time, but if the tool is switched off then the various damping mechanisms will act to quickly bring the vibrations to an end.

Isolation is the process of preventing the transfer of energy from one object to another without necessarily removing energy from the system. A motor vehicle can be considered to be a mass supported on a number of springs. If it is driven over a series of bumps, then the vibration of the wheels will not be transmitted directly to the occupants. The springs themselves will remove very little energy from the system, although damping components will normally also be present. Isolation and damping are considered further in Chapter 16.

Acceleration, velocity and displacement

Three quantities are available which could be used to assess the magnitude of a body's vibration. It was shown earlier that at any instant the vibrating mass will be displaced from the equilibrium position: either upwards, downwards, or – twice in each cycle – it will have zero displacement. This displacement is continually changing, but over a period it will be possible to do what is done for sound pressure: to calculate the rms average of the displacement.

As the displacement is constantly changing, it will at any instant have a velocity, and this may be directed upwards or downwards. Twice in each cycle

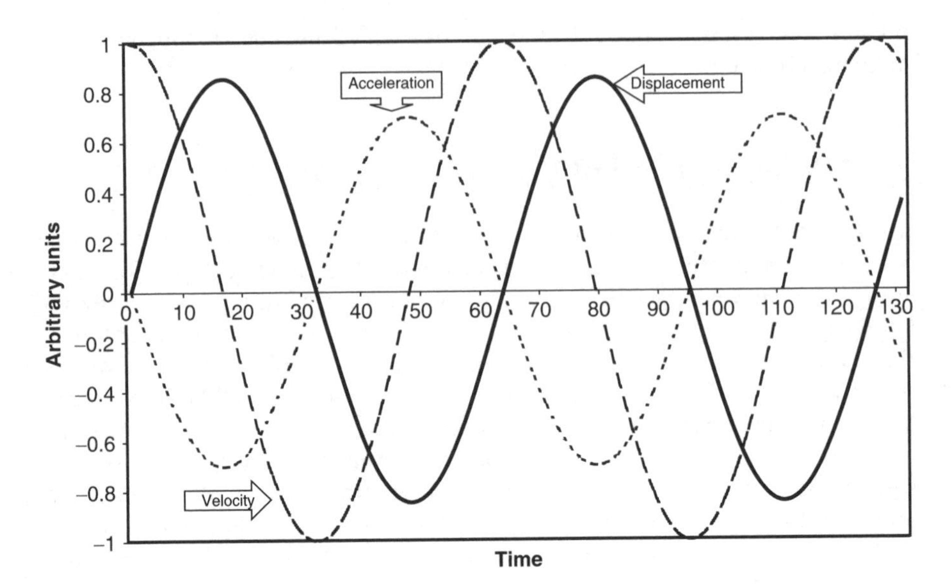

Figure 7.2 Displacement, velocity and acceleration.

the mass will come to a halt as it reaches the maximum displacement at either the top, or bottom of its motion, but this zero velocity is instantaneous and it begins immediately to move back towards equilibrium. A graph of velocity against time has the same shape as the graph of displacement against time. The velocity is continually changing, and a change in velocity is called an acceleration. The acceleration, too changes sinusoidally. Figure 7.2 shows the changing displacement, velocity and acceleration. Although they are plotted here on the same graph, it should be remembered that the amplitudes cannot easily be compared as the three quantities are physically different and are measured in different units.

Displacement is measured in metres (or more likely in millimetres), velocity in metres per second (ms^{-1} or m/s) and acceleration in metres per second squared (ms^{-2} or m/s^2). Because these are different quantities, measured in different units, a direct comparison cannot be made between them. An amplitude of $1.5\,ms^{-2}$ when measuring acceleration, for example, will not normally correspond to an amplitude of $1.5\,ms^{-1}$ when measuring velocity. Although the amplitudes cannot directly be compared, the frequency is the same whichever quantity is measured, and it will be seen that the phase of the three quantities is different. In other words they are shifted sideways relative to each other on the graph. The relationship between acceleration, velocity, displacement and frequency is discussed further in the box.

Vibration magnitudes – whether measured in terms of acceleration, velocity or displacement – can be expressed in decibels. However, this is not normally done and it is a complication which is best avoided.

Of the three quantities, acceleration is the one normally measured, probably because vibration measuring devices normally produce an output proportional to the acceleration. Many of the human responses to vibration seem – at least over part of the frequency range – to be related more closely to velocity. Displacement can be important because it is the quantity that can sometimes be judged visually so it is sometimes easy to check if an apparently high displacement is physically realistic.

Where there is more than one frequency present these quantities can be converted to each other as long as frequency analysis equipment is available.

If the frequency of the vibration is known, then it is not difficult to convert the three quantities to each other. The equations which do this are:

$$v_{rms} = \frac{a_{rms}}{\omega} \tag{7.1}$$

$$x_{rms} = \frac{v_{rms}}{\omega} \tag{7.2}$$

$$x_{rms} = \frac{a_{rms}}{\omega^2} \tag{7.3}$$

where a_{rms} is the rms acceleration; v_{rms} is the rms velocity; x_{rms} is the rms displacement; and ω is 2π times the frequency.

The three axes

Acceleration, velocity and displacement are all vector quantities. To describe them fully, the direction must be known as well as the magnitude. A tool handle can vibrate in one of three directions which are at right angles to each other. It could vibrate from end to end, side to side or up and down. Most likely, its motion will be a combination of all three of these.

In order to get a full picture of its motion it is necessary to make measurements along three axes which are orthogonal; in other words they are at right angles to each other. Traditionally, the three co-ordinate axes used are labelled the x, y and z axes. For hand–arm vibration, and also for whole body vibration, standard axes are defined in relation to the human hand or body. For some purposes, it is sufficient to make an assessment of the vibration magnitude along each of the three axes separately. For hand–arm vibration assessments according to ISO 5349:2001, the values for three individual axes are combined using the three-dimensional form of Pythagoras' equation: details are given in Chapter 11.

Averaging vibration levels over time

Vibration levels are rarely constant for extended periods of time in workplace situations. As with noise, the equal energy principle is used to measure a long-term average for assessing hand–arm vibration levels. Because decibels are not normally used in vibration measurement, the calculations involved in averaging vibration levels over time are slightly simpler than those required to average noise levels.

$$a_{eq} = \sqrt{\frac{a_1^2 \times t_1 + a_2^2 \times t_2}{T}} \qquad (7.4)$$

where a_1 and a_2 are acceleration levels; t_1 and t_2 are the times for which they are each maintained; T is the overall time period (equal to $t_1 + t_2$); and a_{eq} is the continuous vibration level, sustained for the entire period, which would deliver the same quantity of vibrational energy.

An example of this calculation appears in Chapter 11.

Instruments for hand–arm vibration measurement are able to carry out equal energy averaging when – as is normally the case – the variation in acceleration is more complex than in this very simple example.

Whole body vibration exposure can be calculated either by using rms averaging (which assumes that the overall energy involved is the important quantity) or by using a procedure known as root-mean-quad or rmq. rmq averaging lays considerably greater weight than does rms averaging on the high vibration magnitudes, even if they are only sustained for a short period of time. Conversely, the duration of exposure is less important when using rmq averaging than when using rms averaging.

When assessing the daily dose of hand–arm vibration, a quantity known as A(8), is normally calculated. This is the equivalent 8-h level; in other words the steady vibration level which, if maintained continuously for 8 h, would contain the same quantity of energy as was contained in the actual, varying vibration exposure. It is therefore very similar in conception to $L_{EP,d}$, the quantity used in a similar way when assessing noise exposure. The 2001 edition of ISO 5349 renamed A(8) as $a_{hv(eq,8\,h)}$ while recognizing that A(8) would continue to be widely used as the name for this quantity.

Shocks

A shock in vibration terminology has the same meaning as in plain language. 'Jolt' is a word which has a similar meaning in day-to-day language (but 'jolt' is not used in a technical context). Many powered tools rely for their operation on repeated impacts between the tool head and the work. Sometimes this impact action is the only important action taking place; examples of this would be a road breaker or a nail gun. Sometimes, as in the case of a hammer drill, the impacts are superimposed on a rotating motion. In a few cases, the impacts arise out of the action of a rotating tool, as it snatches or jumps during operation.

In terms of analysing vibration amplitude and frequency, shocks will manifest themselves as a high amplitude vibration maintained for a very short period. Where shocks are embedded in a steady vibration level, then frequency analysis would normally show them contributing a significant high frequency component.

8

Hand–arm vibration syndrome

HAVS and vibration white finger

Hand–arm vibration syndrome (HAVS) is the preferred name for the condition caused by prolonged exposure of the hand–arm system to mechanical vibration. It has three components: vascular, neurological and musculoskeletal. The vascular component is the most familiar, and this component is sometimes known as vibration white finger or VWF. The term VWF is in fact sometimes misleadingly used to refer to hand–arm vibration syndrome as a whole. Other names have also been used in the past for the vascular component; 'Raynaud's phenomenon of occupational origin' is a precise but long-winded term, while names used in specific trades include 'dead finger' and 'grinder's cramp'.

The vascular symptoms can have a number of other causes besides vibration exposure. The neurological symptoms, too, can have other causes, although the two sets of symptoms occurring together point more definitely to vibration exposure as the cause. The musculoskeletal component is the component for which the least satisfactory relationships have been established between vibration dose and effects.

The vascular, neurological and musculoskeletal components

The vascular component of HAVS is a form of Raynaud's phenomenon. When exposed to cold, the veins of some of the sufferer's fingers go into spasm, severely restricting the blood supply and causing the affected parts to turn white. At the same time, all feeling in that part of the hand is lost. The attack may last for 30 min, and only finish when the hand is rewarmed. Re-establishment of the circulation is often accompanied by pain.

In mild cases, attacks are infrequent, occurring only when the hands are exposed to severe cold. In these mild cases, the attacks only affect the end joints of one or two fingers. As the condition develops, the attacks become more common, and are triggered by less severe exposure to cold. They affect more joints and spread to more fingers. The thumb and the palm of the hand are rarely affected, but in extreme cases, the disruption of the blood supply to the fingertips can cause the ends of the fingers to become ulcerated and this can, infrequently, lead to gangrene.

The condition is named after Maurice Raynaud, a French physician who first studied and described it in the nineteenth century. He did not connect it with vibration exposure. The condition has a number of causes and he was concerned with what is now called Raynaud's disease. This affects up to 10 per cent of the female and 5 per cent of the male population and has no known cause (in the jargon, it is idiopathic). Other factors which can trigger Raynaud's phenomenon include certain injuries to the hand and arm, the action of certain drugs and poisons, and some medical conditions (Table 8.1).

The neurological component has received more attention over recent years. Less obvious than the vascular component, it may nevertheless interfere more seriously with the sufferer's ability to work and to otherwise lead a normal life. Diagnosis is not easy for either component, but in the case of the neurological component diagnosis is simplified by the fact that there is not a large number of cases with no known cause. The symptoms consist of a progressive numbness which affects increasing numbers of finger joints and may be accompanied by tingling. This numbness may be present continuously, or may be apparent at particular times. However, it is separate from the numbness which accompanies the vasospastic attacks resulting from the vascular component. Numbness and tingling are accompanied by a loss of ability to carry out fine motor tasks, such as manipulating small objects. Sufferers may burn or otherwise injure their fingers as a result of their inability to detect sharp and hot objects. Once again, there are other possible causes of the symptoms (Table 8.2).

The musculoskeletal component includes a range of different conditions affecting the bones and the joints (Table 8.3). None of these conditions is

Table 8.1 Examples of nonvibration causes of Raynaud's phenomenon

No known cause	Genetic factors (?)
Drug treatment	Beta-blockers
Injury to the hand/arm	Frostbite
	Other injuries affecting the vascular and nervous systems in the hand and arm
Diseases	Polio
	Diseases that affect connective tissue such as:
	Scleroderma
	Rheumatoid arthritis
	Systemic lupus erythematosus
	Arterial diseases, such as arteriosclerosis
Exposure to toxins	Ergot
	Vinyl chloride

Table 8.2 Examples of nonvibration causes of neurological symptoms similar to those associated with HAVS

Drug treatment	Streptomycin
	Chloramphenicol
Entrapment of nerves	Carpal tunnel syndrome
Diseases of the central nervous system	Multiple sclerosis
	Spinal cord tumours
Other diseases	Diabetes
Toxins	Organophosphates
	Lead
	Thallium
	Carbon disulphide
	Hexane

Table 8.3 Some musculoskeletal conditions associated with hand–arm vibration

Carpal tunnel syndrome
Kienbock's disease
Osteoarthritis in the thumb
Dupuytren's contracture
Bone cysts (caused by leakage of synovial fluid from the joints)

uniquely caused by vibration exposure. All have an increased incidence among workers exposed to hand–arm vibration, but in most cases it is far from certain that vibration is the cause. All these jobs involve heavy physical work, which is itself likely to cause most of the cases. Vibration exposure may make it more likely that these conditions will develop than in workers doing similar jobs without the vibration exposure. It is not possible, though, to make quantitative predictions about the relationship between vibration exposure and the consequent development of the musculoskeletal component.

The physiological processes underlying the development of HAVS are not well understood. One theory is that the principal damage is to the nervous system, and that this results in a distortion of the normal response to cold. Instead of reducing circulation in the extremities gradually in order to conserve heat, this happens in an exaggerated way. Another theory explains the syndrome in terms of damage to the capillaries in the fingers. They receive the normal nerve signal when the hand is exposed to cold, but are unable to respond in a controlled way. Neither of these mechanisms is satisfactory in explaining all the features of the condition, and a third approach is to assume that the vascular and neurological damage components progress independently, although they interact when a vasospastic attack is triggered.

Diagnosis

Diagnosis of HAVS is hampered by the fact that few objective tests are available. Finger blanching during a vasospastic attack is very visible, but they normally occur early on winter mornings and therefore are unlikely to be seen by a medical practitioner. It is difficult to trigger an attack artificially, and it is generally thought undesirable to do so. The sufferer's report of the frequency and severity of the attacks is the main source of information. Numbness is mainly assessed by interviewing the sufferer about his/her symptoms. Tests are available to quantify the extent of the numbness, but once again they depend on the sufferer's co-operation and objectivity. Those tests which aim at objective confirmation of the vascular component are not very specific to HAVS.

Diagnostic tests for HAVS

Of the various tests available, the following were selected for use to assess eligibility for compensation under the compensation scheme for former British coal miners (Lawson and McGeoch, 2003). There is no universal agreement on the value of any particular set of tests. Some of the tests listed here require specialized equipment which is not widely available.

1. Adson's and Allen's tests supplement blood pressure measurement by providing further information about the blood supply to the hand and arm.
2. Tinel's and Phalen's tests detect impairment of the nerve system in the hand. They are not specific to HAVS but are used in the diagnosis of, for example, carpal tunnel syndrome.
3. The Purdue pegboard test: A test of manual dexterity.
4. Jamar dynamometer test: This measures the strength of the subject's grip.
5. Cold provocation test: Temperature sensors are attached to each finger of a hand which is then cooled by immersion in water. The degree of cooling is not sufficient to provoke a vasospastic attack. After a few minutes the hand is withdrawn from the water and the rewarming process is monitored. A hand affected by the vascular component of HAVS normally takes substantially longer to regain a normal temperature.
6. Thermal aesthesiometry: This is a test of the subject's ability to feel warmth and cold in an affected finger. The finger is placed in contact with a metal disc whose temperature can be accurately controlled, and reported impressions of heat and cold are recorded.
7. Vibrotactile thresholds: This is a test of the subject's ability to detect vibration in a surface in contact with the finger.

It is not possible to assume that sufferers will always act in a co-operative and objective manner when being examined for symptoms of HAVS. Much publicity has been given to a number of compensation claims involving vibration-exposed workers and some claimants are tempted to exaggerate or even invent the symptoms. It has been alleged in the past that potential claimants have been tutored in the symptoms they should report during a medical examination. Conversely, in an organization with a developed health and safety culture, it may be the policy when an employee shows the early signs of HAVS to move them to a job not entailing hand–arm vibration exposure. If this is less well paid work then there is an incentive for sufferers to conceal symptoms.

An experienced occupational health practitioner will take a detailed history of symptoms and of the employee's work history. Where necessary, expert advice can be taken on likely vibration exposure, and the symptoms can be confirmed by the use of the various tests available. An occupational health specialist will probably have experience of checking on fraudulent and exaggerated claims.

The Stockholm workshop scale is used to assess the extent of HAVS (Table 8.4). This assesses the progress of the vascular and neurological components separately, and is carried out separately for each hand (Table 8.5). The Stockholm scale replaced the earlier Taylor–Pelmear scale, which combined the symptoms for the two components. The two components frequently progress at different rates, and the Stockholm scale makes more accurate diagnoses possible. This is important when compensation claims are being processed and an estimate is required of the degree of handicap resulting from the condition. It is also a useful tool when assessing the effectiveness of exposure management programmes.

Another system for assessing the severity of the condition (Griffin, 1990) allocates a numerical score to each finger joint – 1 for the end joint, 2 for the middle joint and 3 for the inner joint. The scores for all the affected

Table 8.4 The Stockholm scale

Stage	Grade	Description
Vascular component		
0		No attacks
1V	Mild	Occasional attacks affecting only the tips of one or more fingers
2V	Moderate	Occasional attacks affecting distal and middle (rarely also proximal) phalanges of one or more fingers
3V	Severe	Frequent attacks affecting all phalanges of most fingers
4V	Very severe	As stage 3 with trophic changes in the fingertips
Neurological component		
0	Vibration exposed but no symptoms	
1SN	Intermittent numbness with or without tingling	
2SN	Intermittent or persistent numbness, reduced sensory perception	
3SN	Intermittent or persistent numbness, reduced tactile discrimination and/or manipulative dexterity	

Table 8.5 Example of an assessment using the Stockholm scale

1L(2)/2R(3) 2SN	Overall assessment. The meaning of each part of this assessment is shown below.
1L(2)	Two fingers of the left hand are subject to occasional blanching attacks affecting only the end joint.
2R(3)	Three fingers of the right hand are affected by occasional blanching attacks affecting the far and middle joints.
2SN	Intermittent or persistent numbness, reduced sensory perception. As no hand is specified, it must be assumed that this assessment applies to both hands.

finger joints are summed to give a number indicating the severity of the case. A single number score is lacking in detail, and although the system has been modified to score each hand – or even each finger – separately, it is not widely used.

Treatment

There is not a great deal that can be done to treat HAVS. There is evidence that the condition of sufferers in the early stages frequently improves by one stage on the Stockholm scale if they are removed from work involving exposure to vibration. At stage one this may mean that the symptoms cease entirely. Removing sufferers from vibration exposure – and doing so as soon as symptoms are confirmed – is therefore probably the single most important step that can be taken to arrest the condition's progress. Existing sufferers from Raynaud's phenomenon from other causes should be identified at pre-employment medical screening and prevented from causing further damage to their blood vessels by the use of vibrating tools.

Although the propensity to vasospastic attacks can be reduced only within certain limits, it makes obvious sense if those who are subject to them avoid the conditions which trigger them. This means wearing gloves and warm clothing, and avoiding work and leisure activities which are likely to expose the hands to cold. This may be difficult to achieve in practice as it could mean a skilled worker moving to less skilled indoor work possibly entailing a reduced income. Certain leisure activities, such as gardening and watching football matches, may also have to be given up altogether as they will often involve exposing the hands to cold. Success has been claimed for the use of biofeedback techniques which train the sufferer to maintain the circulation in the hands even when exposed to cold.

A variety of treatments have been used to alleviate the neurological symptoms, including hot wax treatments. Many of the musculoskeletal symptoms, including carpal tunnel syndrome (CTS) can be successfully treated by surgery.

The relationship between vibration dose and HAVS

The numbers of workers exposed to damaging levels of hand–arm vibration are lower than those exposed to harmful noise levels. Awareness of the problem developed later than was the case with noise, and measuring equipment has become available much more recently. Precise knowledge of the quantitative relationship between noise exposure and hearing loss is inadequate, but it is much more advanced than is the case for hand–arm vibration exposure.

The initial assumption was that, as with noise, an equal energy principle should be used. While this is perhaps not the whole story, it seems to be satisfactory in practice, and it forms the basis of the methods adopted by the ISO in an annexe to the international standard ISO 5349 Part 1:2002. The precise relationship quoted implies that very detailed knowledge is available about the relationship between vibration exposure and the probability of developing HAVS. It predicts the number of years taken for 10 per cent of a given group of employees to develop symptoms of HAVS in terms of their average 8-h equivalent exposure A(8).

$$D_y = 31.8 \times A(8)^{-1.06} \qquad (8.1)$$

where D_y is the average number of years of exposure of the group of employees, in years; A(8) is the average 8-h equivalent exposure of the group, in ms^{-2}, assessed using a root-sum-of-squares procedure.

The effects of the relationship between exposure and the probability of developing HAVS (allowing for a more realistic degree of precision than is implied in equation 8.1) can be summarized as follows:

- An A(8) of 3.0 ms^{-2} continued for 10 years can be expected to cause vascular symptoms in 10 per cent of those exposed.
- A doubling of the 8-h equivalent exposure will approximately halve the time taken for HAVS symptoms to develop.

Table 8.6 shows the number of years taken, according to this standard, for 10 per cent of a population exposed to various daily equivalent levels to develop HAVS symptoms.

Table 8.6 A(8) values and exposure durations which can be expected to cause HAVS in 10 per cent of those exposed

A(8) in ms^{-2}	Exposure time in years
1.0	32
2.0	15
3.0	10
4.0	7
5.0	6
6.0	5

Similar analyses based on the previous version of ISO 5349 and BS 6842:1987 will appear to be in conflict with the relationships described above. This is because they are based on a dominant axis assessment.

There is some evidence that damage to the hands will occur more slowly if the peripheral circulation is poor during vibration exposure. It is normally recommended that those using vibrating tools should keep both their hands and their body warm. Smoking during vibration exposure may increase the damage due to its effect on peripheral circulation, and there is some evidence that short periods of vibration exposure interspersed with vibration-free periods will cause less damage than the same vibration exposure delivered in one continuous period.

The vibration frequencies which cause the most damage have been the subject of some debate. The hand–arm vibration frequency weighting which appears in the ISO standard is based on measurements of human perception of vibration applied to the hand. It does not follow that these are also the frequencies which cause most damage. In the case of noise exposure a frequency weighting based on noise perception has been successfully used to assess the likelihood of hearing damage, but this is not the case with some other physical agents. For example, some forms of radiation can be lethal even though nothing is felt at the time of exposure. The hand–arm frequency weighting attaches most importance to vibration in the range from 8 to 20 Hz, and it is probable that the most damaging frequencies are rather higher than this, say from 25 to 150 Hz. However, the errors introduced by using the existing frequency weighting are probably not serious, and it has not been possible to agree on an improved frequency weighting. The 2001 revision of ISO 5349 made no changes to the existing frequency weighting, which will probably remain in use for some years to come.

9

Vibration measurement

Time averaging

Vibration measurements for use in hand–arm vibration assessments are done on the assumption that the equal energy principle gives valid predictions of the probability of damage occurring. Hand–arm vibration meters will have this assumption built into the software that controls the instrument, and although some are capable of making measurements of instantaneous vibration magnitude, all instruments intended for these measurements also have an equal energy averaging function (comparable to the L_{eq} function on most sound level meters). It is this function that is likely to be used for almost all measurements. The terminology for hand–arm vibration measurements is less advanced than for noise measurement, so that instrument manufacturers use a variety of nonstandard terms on their instrument displays. *Aeq* and *LAeq* are among those in use.

Because the decibel scale is not normally used for vibration assessments, the equations used for averaging multiple vibration exposures is different from – and simpler than – that used for calculating an L_{eq} from subperiod L_{eq}s.

$$a_{hv} = \sqrt{\frac{a_{hv1}^2 \times t_1 + a_{hv2}^2 \times t_2}{T}} \tag{9.1}$$

where a_{hv} is the acceleration averaged over the entire period T; a_{hv1} is the acceleration for the first subperiod t_1; a_{hv2} is the acceleration for the first subperiod t_2; and $T = t_1 + t_2$.

As with noise exposure assessments, the above equation can be modified simply to calculate an 8-h equivalent level, known as A(8).

$$A(8) = \sqrt{\frac{a_{hv1}^2 \times t_1 + a_{hv2}^2 \times t_2}{8}} \tag{9.2}$$

where a_{hv} is the acceleration averaged over the entire period T; a_{hv1} is the acceleration for the first subperiod t_1; a_{hv2} is the acceleration for the first subperiod t_2; and as with noise exposure assessments, a standard 8-h working day is assumed.

It will be seen in Chapter 13 that the equal energy principle is not always used for whole body vibration measurements.

The hand–arm vibration frequency weighting

The frequency weighting used for hand–arm vibration measurements is defined in a number of different standards, and appears in slightly different forms (Figure 9.1). In practice, the differences between the various versions of the frequency weighting are not very important, even though it can be confusing to find apparently different versions, sometimes in the same publication. The different versions arise because the field of hand–arm vibration exposure assessment is relatively new. Tentative relationships established from research data are most easily represented as straight line graphs. When standards are developed specifying acceptable performance from a measuring instrument, frequency weightings must be put into a form which can easily and economically be

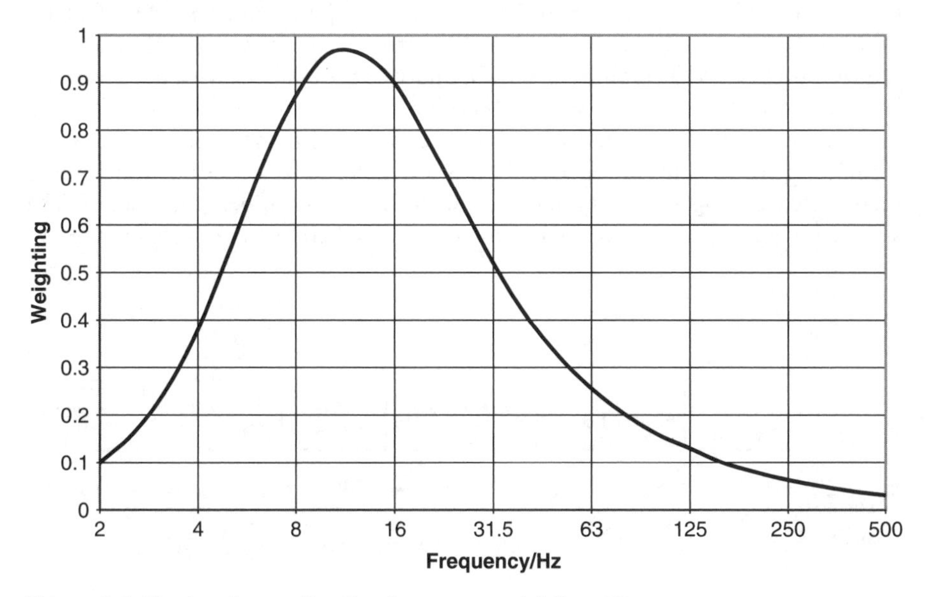

Figure 9.1 The hand–arm vibration frequency weighting, W_h.

achieved by the electronic engineers who design the frequency weighting filters. This is almost impossible with straight line frequency responses, and at this stage the straight line relationships need to be converted to a form which is attainable using one of the standard designs of electronic filter. Since measuring equipment needs to be checked against the specification in the relevant instrumentation standard – in this case ISO 8041 Part 2 – this is most usefully treated as the definitive version of the frequency weighting.

All instruments intended for hand–arm vibration measurement will have this frequency weighting built in. ISO 8041:1990 requires them also to have a linear frequency weighting. This can be useful for calibration purposes, but would give wrong results if used for measurements. Frequency analysis equipment is occasionally used in hand–arm vibration work, and Appendix A shows the frequency weighting in a form that can be used to calculate weighted acceleration values from a frequency analysis.

Vibration meters

Although there are many vibration meters on the market, only those designed specifically to measure hand–arm vibration can be used to measure HAV exposure. Many of these are designed just for HAV measurement, while others have the capability to make whole body vibration measurements too. One or two instruments are on the market which can be used alternatively or simultaneously as sound level meters.

Outwardly, many HAV meters look very like sound level meters – some are built inside an identical case (Figure 9.2). The microphone of a sound level meter is normally mounted on the front of the instrument because the measuring position is close to the employee's head, and it is normally easy to hold a sound level meter in this position while observing the display. Hand–arm vibration measurements, though, must be made in contact with the tool handle and it would be inconvenient for the whole vibration meter to be at this position, even if there were no danger of this interfering with the vibration being measured. The accelerometer – the sensor of a vibration meter – is therefore connected to the rest

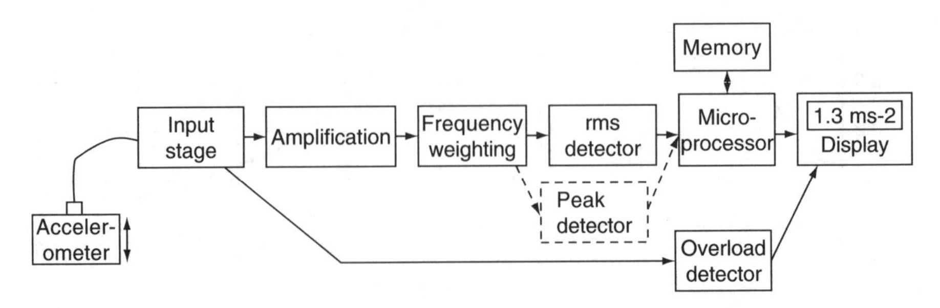

Figure 9.2 Block diagram of a hand-arm vibration meter.

of the measuring instrument by a short length of cable, typically between 0.5 and 1 m long, although it may be possible to use a longer cable if this is required for the particular measurement being carried out.

Accelerometers

Most accelerometers used for HAV measurement are piezoelectric types. Piezoelectric materials have the property that when they are compressed, a voltage is generated across the opposite faces of the crystal. These materials are also used in many gas lighters. In the simplest form of piezoelectric accelerometer, a piezoelectric crystal is arranged so that it is alternately compressed and stretched by the movement of the vibrating body, and the electrical impulses created as a result are passed to the input of the vibration meter (Figure 9.3).

Practical accelerometers are normally configured so that the piezo material is subject to shear deformation rather than compression (Figure 9.4).

Other types of accelerometer are available, including piezoresistive, capacitative, and laser interferometery types, but piezoelectric accelerometers are almost always used for hand–arm vibration measurement.

Clearly a useful accelerometer will operate in a linear fashion when exposed to vibrations with the expected range of accelerations. That is to say that a doubling of the acceleration should lead to a doubling of the output. They will do this for a limited range of frequencies. Figure 9.5 shows the frequency response of a typical accelerometer. The useful operating frequency range is the range over which the output is independent of frequency, and it is limited by the resonant frequency of the accelerometer. This needs to be significantly larger than the highest frequency to which the accelerometer is expected to be exposed. However, the frequency response as measured in a laboratory assumes a rigid connection between the accelerometer and the vibrating surface. In practice, when attached to a tool

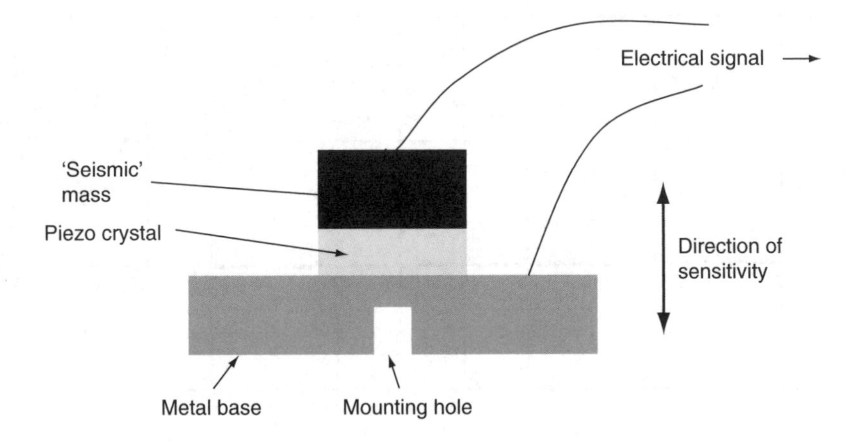

Figure 9.3 Schematic diagram of a compression accelerometer.

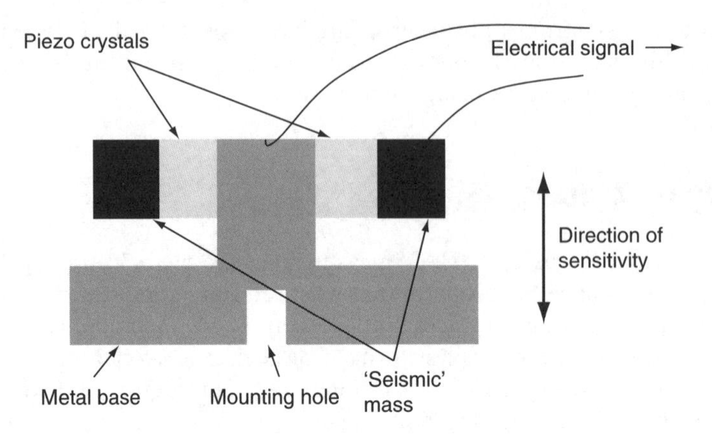

Figure 9.4 Schematic diagram of a shear accelerometer. In practice three seismic masses are arranged around a triangular central post.

handle the connection to the vibrating surface will be less than perfect (as discussed later in this chapter), and on occasion it may be very poor. This has the effect of reducing the resonant frequency and hence also the useful frequency range of the accelerometer. It is therefore important to use an accelerometer with a resonant frequency at least four times as high as the highest frequency present (Figure 9.5).

The output from an accelerometer is a high impedance one which needs to be converted to a lower impedance by a preamplifier. Two types of preamplifier have been commonly used:

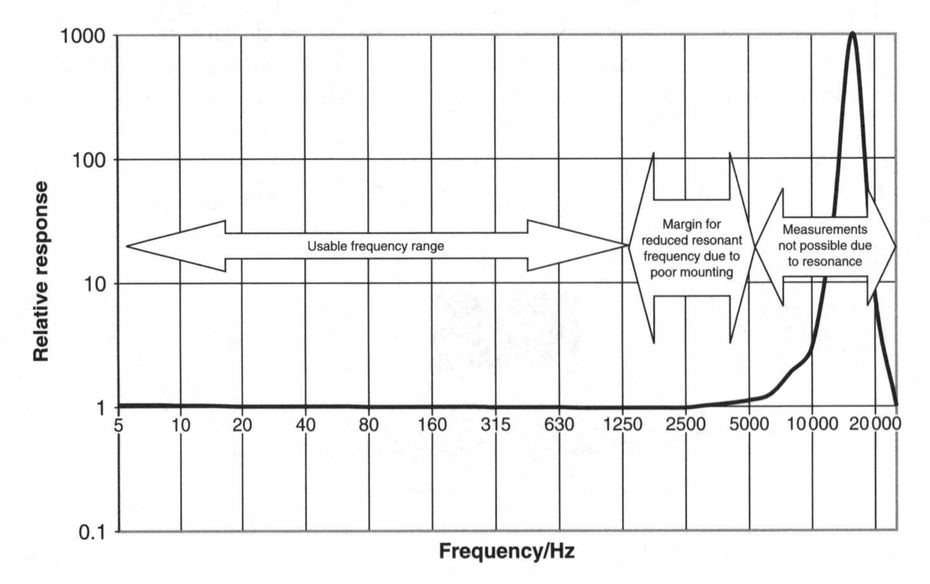

Figure 9.5 Frequency response of an accelerometer.

- A voltage preamplifier outputs a voltage proportional to the voltage from the accelerometer.
- A charge preamplifier outputs a voltage proportional to the charge flowing through the accelerometer.

Charge preamplifiers are nowadays considerably more common. One advantage they have is that they are relatively insensitive to the use of different lengths of cable between the accelerometer and the preamplifier. Because two different types of preamplifier are used, two sensitivities are normally specified in the accelerometer's data sheet:

- A charge sensitivity measured in picocoulombs per metre per second squared (or pC/ms^{-2}), and
- A voltage sensitivity measured in millivolts per metre per second squared (mV/ms^{-2}).

One of these values – normally the charge sensitivity – will be needed when entering the sensitivity to calibrate the vibration meter. Alternatively, if calibration is carried out by means of a fixed-level calibrator, then it can be used to check that the sensitivity has not shifted. The sensitivity will also be important when considering using a different accelerometer.

$$\text{sensitivity} = \frac{V_{rms}}{a_{rms}} \tag{9.3}$$

where V_{rms} is the rms voltage output by the accelerometer and a_{rms} is the rms acceleration to which it is exposed.

Many accelerometers now have a miniature electronic preamplifier built into their casings, and receive a power supply via the same cable which carries the signal. These types of accelerometer are sometimes used with hand–arm vibration meters, and in this case a voltage amplifier input is needed on the vibration meter itself. Therefore a hand–arm vibration meter designed for use with this type of accelerometer will not normally be compatible with the standard type of accelerometer and vice versa. These accelerometers have one major disadvantage in that the built-in preamplifier has to be capable of handling the entire range of input signals from the accelerometer. Although electronic technology advances very fast, this has in the past been an unrealistic requirement for these very small preamplifiers so that they are prone to overloading when used to measure vibration with an impulsive component. The sensitivity of an accelerometer is normally related to its physical size, with smaller accelerometers being less sensitive. While it is often desirable to use as small an accelerometer as possible this will be limited, among other factors, by the need for an easily measurable output voltage.

Most accelerometers can only measure along one axis. A triaxial accelerometer consists essentially of three accelerometers mounted inside one case, and oriented along mutually perpendicular axes. They will clearly be bulkier than a single axis accelerometer, and this can be a problem when mounting these accelerometers on a tool handle. To minimize this problem the individual accelerometers need to be very small, and this can lead to problems with low sensitivity. On the other hand, if they can be used, and if a three-channel vibration meter is available, they are likely to cut measurement time by a factor of three.

Mounting accelerometers on tool handles

The ideal way to mount an accelerometer on a tool handle would be to grind flat a small section of the handle, and then to drill and tap a hole to take the standard threaded mounting stud. Although this may occasionally be an option when testing prototype tools, normally it is completely out of the question.

For hand–arm vibration measurements the most common mounting method is to use a small aluminium block which has a threaded hole in three of its sides, each aligned along a different axis. The accelerometer can be mounted in turn along each of the three axes, although if a triaxial accelerometer is used only one mounting hole is required.

The mounting blocks (Figure 9.6) can be attached to the tool handle in a number of ways:

- A V-shaped block with a slot through which cable ties can be passed to strap it to the handle;
- A T-shaped block which is held between the operator's fingers;
- A flat or curved bar held under the operator's palm; the mounting block is attached to one end on this.

All of these variations have disadvantages:

- Cable ties need to be very tightly fastened, normally using a tool to tighten them.
- Between-the-fingers blocks sometimes have a long extension, raising the possibility that they will pivot on the tool handle and the accelerometer will be undergoing vibration with a considerably greater magnitude than is present at the tool handle (this is sometimes called the cantilever effect).
- Similar blocks without the extension are very awkward to hold.
- The flat bar types sometimes can only accommodate one or two accelerometer mounting holes so that it is difficult to measure all three axes.

Since most mounting blocks will be difficult to use in some situations, it is best to have a variety of types available.

Other mounting methods may be possible. The accelerometer can be attached directly to the tool handle using a cyanoacrylate or epoxy adhesive, although recovering the accelerometer and removing adhesive residues from the tool handle are obvious problems. If the accelerometer is screwed to a threaded stud

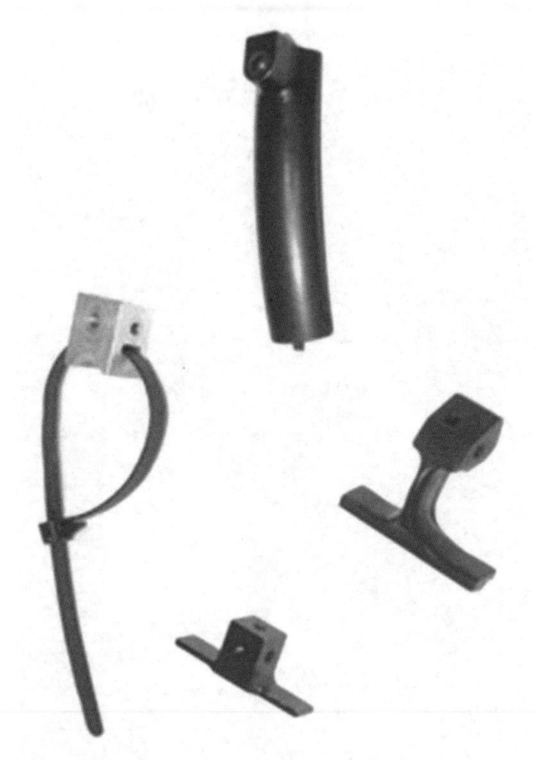

Figure 9.6 Mounting blocks.

which is glued to the tool handle, then the accelerometer at least is reusable even if the tool handle is likely to be permanently affected. A small dab of softened beeswax is often used to attach accelerometers to surfaces for other vibration measurements. The surface needs to be clean and dry, but – more seriously for tool handle measurements – this method is not rugged enough to stand up to physical contact with the operator's hand or other objects.

A large hose clip can sometimes be fitted round a large tool handle. This can be tightened to ensure a rigid connection between the accelerometer and the tool handle. If a screwed stud is preglued to the hose clip, then the accelerometer can be easily attached. However, fastening the hose clip to the handle of each tool can be a lengthy process, and a number of different sized hose clips are needed to cope with the range of tool handles encountered.

Calibration

Vibration calibrators are similar to sound level calibrators in so far as they are devices which impose a known acceleration when an accelerometer is correctly

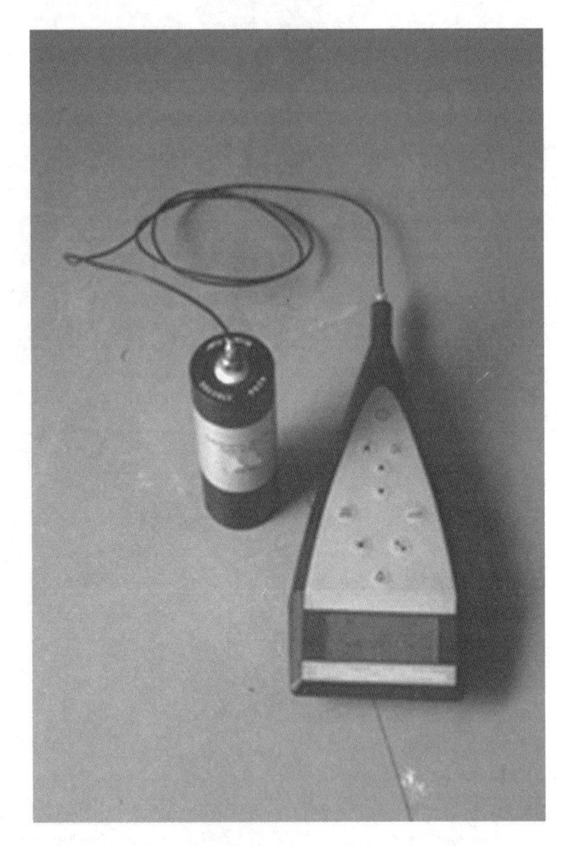

Figure 9.7 Calibration of a hand–arm vibration meter.

attached (Figure 9.7). Both portable and laboratory calibrators are in use. If not acquired as part of a set with the accelerometer and vibration meter, care needs to be taken that they are compatible. The accelerometer must have a mass which falls within the range for which the calibrator is able to generate its fixed acceleration.

The acceleration generated by the calibrator will not be the same as the figure displayed by the instrument unless calibration is carried out using linear frequency weighting. Instruments will normally have a linear weighting available as is required by the standard. To avoid switching frequency weightings, the rated calibrator acceleration can instead be multiplied by the W_h frequency weighting at its operating frequency to calculate the acceleration to be displayed when the calibrator is used and the instrument is using the W_h frequency weighting. These factors can be found in Appendix A.

Example

A calibrator which generates 6.92 ms^{-2} at 80 Hz is used with a hand–arm vibration meter. What reading is expected if the W_h frequency weighting is selected?

The frequency weighting at 80 Hz is 0.2, so the acceleration is $6.92 \times 0.2 = 1.38$ ms^{-2}.

Although electronic instruments such as vibration meters are nowadays extremely stable, calibration provides a useful check on the function of the measuring chain as a whole. The cable connecting the accelerometer to the meter is vulnerable to damage, and field calibration would make it clear that this was faulty if this was not obvious for other reasons. ISO 5349 Part 2 requires field calibration to be carried out.

The issues involved in periodic verification of sound level meters are discussed in Chapter 3 of this book. A number of laboratories offer a periodic verification service for hand–arm vibration meters. There is at present no standard which defines the tests to be carried out, so a reputable laboratory will derive a set of tests from a standard such as ISO 8041, and will specify exactly which tests have been carried out in the test report (and preferably also in their quotation for the work). Because vibration calibrators are not always available, some organizations may offer a service which is essentially a field calibration carried out in the laboratory. This should be made clear when they offer the service, and will be reflected in the price charged.

Vibration calibrators will also need to be verified, either at the same time as the HAV meter or more frequently.

dc shift

Piezoelectric accelerometers sometimes suffer from a problem known as dc shift. This occurs when they are exposed to vibration with a strong impulsive component such as is experienced when using impact tools, such as chipping hammers. A dc voltage is superimposed on the expected ac voltage for a period which is short in absolute terms but covers several periods of the vibration. As a result, the rms voltage as detected by the vibration meter is extremely high, and this will be interpreted as a very high acceleration value. This does not happen with other types of accelerometer, and advice on dealing with the problem if it occurs can be found in Chapter 12.

10

The development of legal controls on hand–arm vibration exposure

Early work

Hand–arm vibration syndrome (HAVS) is a condition which is caused by prolonged exposure of the hand/arm system to continuous vibration. It is essentially a modern disease. There are few reports of conditions which can be identified with HAVS before the start of the twentieth century. One of the first studies to describe the condition and its causes was carried out by Dr Alice Hamilton, a physician working for the US Department of Labor in 1917–1918. She studied the effects on workers in stone quarries in Indiana and described the symptoms and discussed the relationship between the cause of the condition (vibration exposure) and the trigger for attacks of 'white finger' (exposure to cold) in a statistical digest published by her department later in 1918. Other important work was carried out in Italy.

Later, many US occupational physicians were surprisingly reluctant to admit that the condition existed. It began to affect large numbers of forestry workers in the 1960s after small, hand-held chain saws were introduced in the 1950s. Up to that time, a chain saw was a much larger device used mainly for felling trees. The smaller saws could also be used to remove side branches, so that the operators were exposed to vibration for a much longer period during the day. It was the subject of large-scale studies in the UK, northern and eastern Europe, and the Far East during the 1960s and 1970s.

Table 10.1 Tools and industries for which HAVS is a prescribed industrial disease

Tool/process	Industry
Chain saws	Forestry
Grinding, sanding or polishing tools	Metal working
Percussive tools used for riveting, caulking, chipping, hammering, fettling or swaging	Metal working
Hand-held percussive hammers	Quarrying, demolition, road repair and construction
Leather pounding machines	Shoe manufacture

ISO 5349:1986, BS 7842:1987 and HS(G)88

By the early 1970s, the condition was well known to be caused by exposure to a number of industrial processes (for example see Kinnersley, 1973). Work continued on agreed procedures for assessing the progress of the disease, and for measuring hand–arm vibration levels. In the UK, the 1974 Health and Safety at Work Act increased the responsibility of employers for the health of their employees, and a number of present and former employees succeeded in suing their employers during the 1970s and 1980s for compensation. An international standard, ISO 5349, was introduced in 1989 for the measurement of hand–arm vibration and British Standard BS 6842 was published in 1987.

BS 6842:1987 was not identical with the international standard. The assessment of hand–arm vibration exposure was (and still is) a new and fast-developing subject. The British Standard committee were not convinced that all the information included in the annexes to ISO 5349:1989 was well founded, and they were not impressed with the advice on assessment procedures. Meanwhile, the consensus in the UK was moving in favour of making HAVS a prescribed industrial disease, and this was done in 1985 (Table 10.1).

Detailed guidance on the management of hand–arm vibration was published by the Health and Safety Executive, as the booklet HS(G)88 Hand–Arm Vibration (1994). By this time, practical equipment was on the market for carrying out hand–arm vibration measurements in the field. The guidance drew on British Standard BS 6842:1987 for recommended procedures for measuring and assessing hand–arm vibration exposure. It recommended that employers should take certain actions if the exposure of any employee to hand–arm vibration was such that the value of A(8) exceeded 2.8 ms^{-2}. Using BS 6842:1987, this was assessed by measuring vibration magnitudes along three axes, but then discarding all but the highest of the measured single-axis values. The actions recommended by HS(G)88 when this A(8) level is exceeded are:

- Provision of training and information for exposed employees
- Measures to control the vibration exposure of employees by substituting lower vibration tools and by working to reduce the grip forces used

- Action to help employees in maintaining blood circulation during vibration exposure, including the provision of warm clothing and the selection of appropriate tools
- A programme of health surveillance.

This publication did not recommend a maximum permissible hand–arm vibration exposure. It did point out that the recommended action level was not one which would eliminate all risk of vibration damage

In the UK, attention was focused on hand–arm vibration during the years following the publication of HS(G)88 by a number of well-publicized civil court cases. *Armstrong v British Coal Corporation* and *Hall v British Gas* (1998) are cases which established that employers should have recognized the potential health risks from hand–arm vibration and taken action to limit them by the mid-1970s (Carling, 1999). The insurance industry was quick to spot the potential liabilities involved, and enforcing authorities prioritized the hazards arising from hand–arm vibration exposure in response to the work leading up to the issue of the Physical Agents (Vibration) Directive. Concurrently with this work, international standards bodies were working on a revised version of ISO 5349 which would take account of new research information and of the needs of those who would be responsible for implementing and enforcing new legislation.

Also during the 1990s, standards and legislation were being developed which would make much more information available to employers who were trying to bring their workforce's hand–arm vibration exposure under control. The 1989 Machinery Directive, implemented in the UK as the Supply of Machinery (Safety) Regulations 1992, required manufacturers and importers to measure and declare hand–arm vibration magnitudes at tool handles. In order for this information to be of any use, the conditions under which measurements are made on different tools of the same type need to be standardized. The ISO 8662 series of standards establishes standard test conditions for the measurement of vibration at the handles of a wide range of tools – mainly pneumatically powered ones – used in the engineering industry. Other standards perform a similar job for other types of tool, notably electrically powered tools and those commonly used in agriculture, forestry and horticulture.

ISO 5349:2001 Parts 1 and 2

By the end of 2001 the International Standards Organization had published a new edition of ISO 5349. Widely agreed to be a great improvement on the earlier standard, this was split into two parts. ISO 5349 Part 1 deals with the principles of hand–arm vibration assessment, while one of its annexes discusses the scientific evidence available on the dose–effect relationship. Part 2 deals with detailed protocols for the measurement of hand–arm vibration. Both standards were quickly adopted as British Standards, BS EN ISO 5349 Parts 1 and 2.

One of the main changes in the new edition of ISO 5349 was the use for assessing HAV exposure of quantities calculated by a root-sum-of-squares combination of the values measured on the three axes – this combined value is sometimes called the vector sum. The combined axes values are inevitably higher for any given measurement than the values previously used which simply consisted of the highest of the three figures from the individual axes. Thus, for consistency, any action level in use based on the previous assessment technique will be equivalent to a higher value when assessed using combined axes. The actual difference between the two figures can vary, in theory, from zero (when all the vibration energy is concentrated along one axis) to the situation where the magnitudes along all three axes are approximately equal and the combined value is about 70 per cent higher than each of the individual axis figures. Measurements on actual tools and processes tend to fall between these extremes, and typically, averaged over a number of different tools, the combined axis values tend to be about 1.4 times the highest axis value.

For example, a value of 2.8 ms^{-2} according to an assessment using the highest of the individual values measured along the three axes would correspond to a value of around 4.0 ms^{-2} when combining data from all three axes.

The new ISO 5349 not only replaced the earlier version of the international standard, but was also adopted in the UK as a British Standard to replace BS 6842:1987. The guidance in HS(G)88 was based on this latter standard which had now been withdrawn. The Health and Safety Executive made some minor amendments to this publication, but decided that a new edition should be delayed until after the publication of the UK regulations to implement the Physical Agents (Vibration) Directive. Although this meant that there would be an interim period during which official advice would apparently conflict with current standards and European legislation, it was judged that in practice an employer complying with the advice in HS(G)88 would also be in compliance with any new guidance.

The Physical Agents (Vibration) Directive

The perceived need for legislation throughout the European Union on HAV exposure was one of the driving forces behind the Physical Agents Directive when this was first proposed in the early 1990s. It became clear by the end of that decade that a directive on the scale originally envisaged was unlikely to be realized. Many of those concerned saw vibration as the key physical agent requiring legislation, and early in 1999 the German presidency proposed moving ahead with a directive covering vibration alone. The idea was not that work on the other physical agents should be abandoned, but that progress was more likely if they were covered by separate directives.

Late in 2000 the European Commission and council agreed on the form of a directive to put before the European Parliament. This finally passed through the

Table 10.2 Duties under the Physical Agents (Vibration) Directive

Duty	When applicable
Consultation with employees	Whether or not exposure exceeds the action value
Provide information and training to employees	When exposure exceeds the action value
Carry out health surveillance	When exposure exceeds the action value
Assess the vibration exposure of employees	Whether or not exposure exceeds the action value
Eliminate or minimize health risks	Whether or not exposure exceeds the action value
Reduce employee vibration exposure to a minimum	When exposure exceeds the action value
Reduce vibration exposure below the exposure limit value	When exposure exceeds the limit value
Take steps to ensure vibration exposure does not exceed the exposure limit value	When exposure exceeds the limit value

remaining stages by June 2002. Like other directives it allowed a 3 year period for member states to incorporate it into their domestic legislation.

The Physical Agents (Vibration) Directive establishes an exposure action value at an A(8) value of 2.5 ms^{-2} and an exposure limit value at an A(8) of 5.0 ms^{-2}. In both cases, the 2002 version of ISO 5349 is to be used to assess employee exposure. Employer duties when one of these values is exceeded are shown in Table 10.2. At the time of writing, the UK regulations have not been finalized, and the wording may differ from that in the directive, although they are required to ensure a level of protection to employees which is no less than that in the directive itself.

Because of the equal-energy averaging used, the difference between an action level of 2.8 ms^{-2} and an action level of 2.5 ms^{-2} is greater than it looks. When setting time limits on the use of a particular tool, the time allowed before the lower limit is exceeded is about 25 per cent less than that which would be set for an action level at 2.8 ms^{-2}. More important is the move from highest-axis to root-sum-of-squares assessment. It has been argued that the exposure action value of 2.5 ms^{-2} is a significant reduction compared to the earlier action level which would be equivalent to an action level of 4.0 ms^{-2} using the new ISO 5349 assessment procedures. It has conversely been argued that employers who, in the absence of a defined limit to exposure, tried to keep exposure below the equivalent of 4.0 ms^{-2} will see the new exposure limit value of 5.0 ms^{-2} as a relaxation of previous guidance.

In practice, the new action and limit values together do not represent either a significant tightening or relaxation of standards. However, the greater clarity resulting from the new directive's requirements will make it easier for enforcement authorities to intervene when they feel justified in doing so, and this is a surer way of reducing health risks than simply altering the numerical limits.

11

Calculating hand–arm vibration doses and limits

Introduction

In this chapter the various calculations which may be required during or after an assessment of hand–arm vibration exposure are collected together. For each type of calculation, a worked example is included and further examples can be found in the Appendix.

The various calculations which may be required are:

- Calculation of the root-sum-of-squares (rss) acceleration from three individual axis measurements
- Calculation of the 8-h equivalent level A(8) due to the use of a single tool
- Calculation of the 8-h equivalent level A(8) due to the use of more than one tool
- Calculation of a time limit on the use of a particular tool
  - To keep exposure below the exposure action value
  - To keep exposure below the exposure limit value
- Calculation of a time limit on the use of a particular tool when another tool has already been used on that day for a known period.

Combining measurement data from three axes

Thus the measurements from the three axes are to be combined by the root-sum-of-squares (rss) method.

$$a_{hv} = \sqrt{a_{hwx}^2 + a_{hwy}^2 + a_{hwz}^2} \tag{11.1}$$

where a_{hv} is the combined vibration magnitude; a_{hx}, a_{hy} and a_{hz} are the single axes acceleration magnitudes.

> ***Example***
>
> Measurements at the handle of a percussion drill show hand–arm weighted accelerations of 1.2, 6.4 and 9.7 ms^{-2} for the x, y and z axes. What is the overall acceleration level according to ISO 5349:2001?
>
> $$a_{hv} = \sqrt{a_{hwx}^2 + a_{hwy}^2 + a_{hwz}^2} = \sqrt{1.2^2 + 6.4^2 + 9.7^2} = 11.7\,\text{ms}^{-2}$$

The chart published by the Health and Safety Executive (1994) can also be used to calculate A(8) values. Figure 11.1 shows how this is done.

The previous standards for carrying out hand–arm vibration assessments, ISO 5349:1989 and BS 6842:1987 allowed the use of the highest of the three axis values rather than a root-sum-of-squares (rss) combination of the three.

Using the new standard, measurements on the same tool can be up to 70 per cent higher than a previous assessment which may have used the earlier standards. On average, measurements using the current standards can be expected to come out about 40 per cent higher than previous ones using the old standards.

A(8) calculation from a single period of exposure

If an employee operates the same machinery under the same conditions for the whole of an 8-h shift, then the vibration magnitude which results from the above calculations will be the same as A(8), the 8-h energy equivalent acceleration. In practice things are never so simple, and some account has to be taken of the time for which an individual is exposed to significant vibration levels in the course of a working day.

In the simplest case, the exposure is to one acceleration level, but it lasts for less than a full 8-h shift. In this case:

$$\text{A}(8) = a_{hv} \times \sqrt{\frac{t}{8}} \tag{11.2}$$

where a_{hv} is the acceleration level to which the employee is exposed; and t is the period of exposure.

In fact, the same calculation would work in the event that the exposure time was greater than 8 h in the course of a day.

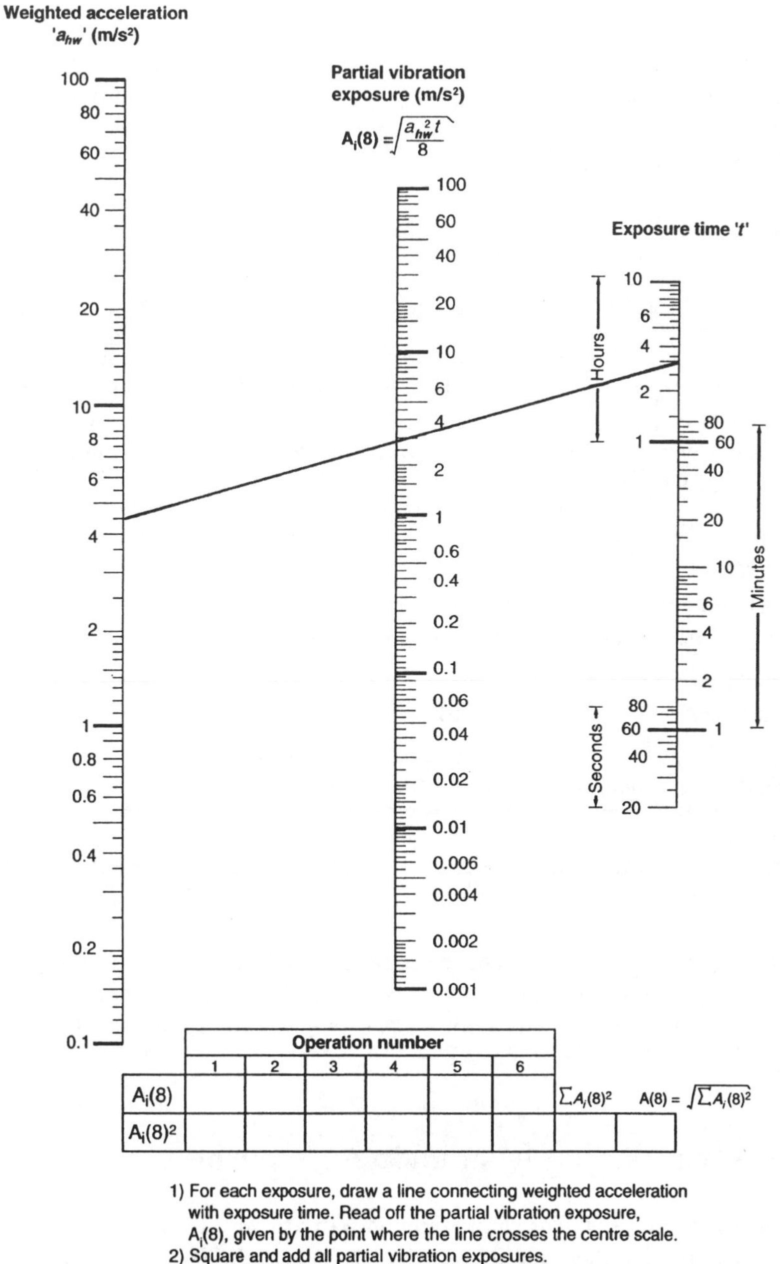

Figure 11.1 A(8) calculation using the HSE chart. From HS(G) 88: Hand Arm Vibration (HSE, 1994). © Crown copyright material is reproduced with the permission of the Controller of HMSO and Queen's Printer for Scotland.

Example

An employee uses a hand tool for approximately 3 h a day, and an acceleration of 4.5 ms^{-2} is measured at the handle while it is in use. Estimate the 8-h exposure level.

$$A(8) = a_{hv}\sqrt{\frac{t}{8}} = 4.5 \times \sqrt{\frac{3}{8}} = 2.76\,ms^{-2}$$

The 1989 version of ISO 5349 (but not BS 6842:1987) used the quantity A(4) to assess daily exposure instead of A(8). This normalized exposure to a notional 4-h day instead of an 8-h one. If earlier assessments are expressed in terms of A(4) they can be converted to A(8) values for comparison purposes by multiplying the A(4) value by 0.707 (i.e. $\frac{\sqrt{2}}{2}$).

Example

Measurements at a tool handle lead to an A(8) of 3.1 ms^{-2}. Is this compatible with an earlier measured A(4) value of 4.2 ms^{-2}?

$$A(8) = A(4) \times \frac{\sqrt{2}}{2} = 4.2 \times 0.707 = 2.97\,ms^{-2}$$

This is within the normal range of uncertainty.

Figure 11.1 shows how to use the nomogram from the 1994 Health and Safety Executive publication *Hand–Arm Vibration* to work out the 8-h equivalent level A(8) in the earlier example of a tool used for 3 h per day and exposing its user to an acceleration of 4.5 ms^{-2}. The acceleration level of 4.5 ms^{-2} on the left-hand column is joined to the exposure time of 3 h on the right-hand column. The line crosses the middle column at a point corresponding to the equivalent 8-h exposure, just under 3 ms^{-2} in this case.

A(8) calculation from multiple exposures

Frequently a worker will use more than one vibrating tool in the course of a single shift. In that case the equation used to calculate the A(8) value needs to be extended to take account of all the significant exposure periods.

$$A(8) = \sqrt{\frac{a_{hv1}^2 \times t_1 + a_{hv2}^2 \times t_2 + a_{hv3}^2 \times t_3 \ldots}{8}} \tag{11.3}$$

where a_{hv1}, a_{hv2}, a_{hv3}, etc., are the a_{hv} values measured during each period of significant vibration exposure; t_1, t_2, t_3, etc., are the time periods over which each of these levels was maintained.

Example

An employee uses hand tools to do two different jobs:

Task	Duration	A(8)
Grinding	4 h/day	2 ms^{-2}
Drilling	2 h/day	3.5 ms^{-2}

What will be the 8-h exposure of this employee?

$$\mathrm{A(8)} = \sqrt{\frac{a_{hv1}^2 \times t_1 + a_{hv2}^2 \times t_2 + a_{hv3}^2 \times t_3 \ldots}{8}} = \sqrt{\frac{2^2 \times 4 + 3.5^2 \times 2}{8}} = \sqrt{\frac{40.5}{8}}$$

$$= 2.25\,\mathrm{ms}^{-2}$$

Time limits

In many cases it will be necessary to calculate a time limit which will enable an employee's daily hand–arm vibration dose to be kept below the exposure action or limit value. Once again, this is simplest when only one tool is in use which involves exposure to significant hand–arm vibration.

$$t_{\max} = \left(\frac{\mathrm{A(8)}}{a_{hv}}\right)^2 \times 8\,\mathrm{h} \tag{11.4}$$

where $t_{\max}$ is the maximum time limit to be imposed; A(8) is the 8-h equivalent exposure limit; a_{hv} is the hand–arm weighted acceleration at the tool handle.

Example

Measurements of a_{hv} at the handle of a shaft grinder during finishing operations yielded a value of 9.4 ms^{-2}. For how long can an employee use this machine before the 8 h equivalent level A(8) reaches 2.5 ms^{-2}? How would this limit be affected if the employer decided to apply a more stringent A(8) limit of 2.0 ms^{-2}?

For an A(8) limit of 2.5 ms^{-2}:

$$t_{\max} = \left(\frac{\mathrm{A(8)}}{a_{hv}}\right)^2 \times 8 = \left(\frac{2.5}{9.4}\right)^2 \times 8 = 0.57\,\mathrm{h}\ \text{or}\ 34.0\,\mathrm{min}.$$

(*Continued*)

Example *(Continued)*

For an A(8) limit of 2.0 ms^{-2}:

$$t_{\max} = \left(\frac{A(8)}{a_{hv}}\right)^2 \times 8 = \left(\frac{2.0}{9.4}\right)^2 \times 8 = 0.36\,\text{h or } 21.7\,\text{min.}$$

A 20 per cent reduction in the exposure limit leads to a nearly 40 per cent reduction in the time limit.

Figure 11.2 shows how to use the HSE chart to determine an exposure time limit. In this case the previous example is used to derive the time limit for a worker exposed to an acceleration level of 9.4 ms^{-2} and whose exposure it is required to keep below the exposure action value of 2.5 ms^{-2}. A line joining 9.4 ms^{-2} on the left-hand column to 2.5 ms^{-2} on the middle column is extended to show on the right-hand column the resulting time limit of just over 30 min.

Table 11.1 shows some example time limits to keep within the exposure action and limit values, and similar information can be derived from Figure 11.3.

Residual time limits

Occasionally it may be necessary to set a time limit for an employee's use of a particular tool on the basis that some exposure to hand–arm vibration has already taken place that day. Trial and error, using the A(8) calculations demonstrated earlier, is one way to calculate this time limit. A more systematic approach uses the following equation:

$$t_2 = \frac{8 \times A(8)^2 - a_{hv1}^2 \times t_1}{a_{hv2}^2} \qquad (11.5)$$

where an exposure has already taken place to an acceleration of magnitude a_{hv1} lasting for time period t_1; a_{hv2} is the acceleration at the handle of the tool to be used next; and t_2 is the maximum permissible time for which it can be used; A(8) is the A(8) value within which it was required to keep.

Example

A worker who has already spent 2 h using a tool with a vibration magnitude of 3 ms^{-2} is now required to use a tool with an a_{hv} of 4ms^{-2}. What time limit should be imposed on this second tool in order to keep the daily exposure A(8) below 2.5 ms^{-2}?

$$t_2 = \frac{8 \times A(8)^2 - a_{hv1}^2 \times t_1}{a_{hv2}} = \frac{8 \times 2.5^2 - 3^2 \times 2}{4^2} = 2\,\text{h}$$

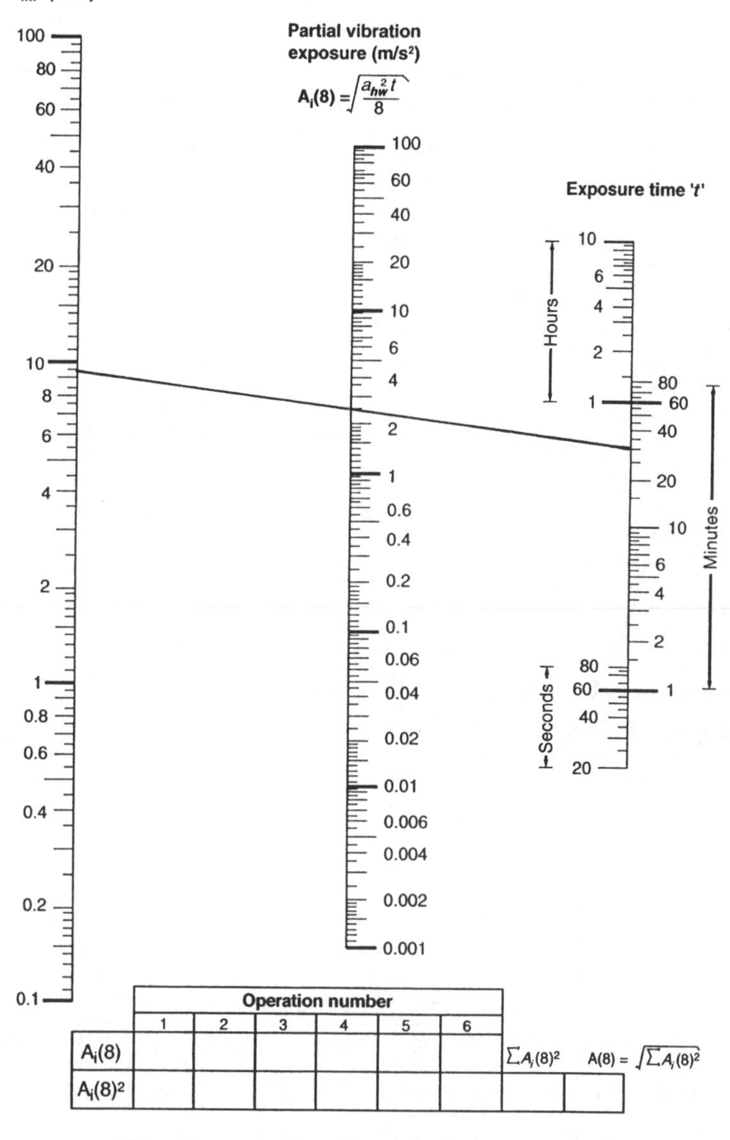

Figure 11.2 Time limit calculation using HSE chart. From HS(G) 88: Hand Arm Vibration (HSE, 1994). © Crown copyright material is reproduced with the permission of the Controller of HMSO and Queen's Printer for Scotland.

Table 11.1 Time limits to keep below Physical Agents Directive action and limit values

a_{hv}/ms^{-2}	Time limit to keep within	
	EAV	ELV
1.5	22 h	No limit
2.0	12½ h	No limit
2.5	8 h	No limit
3.0	5½ h	22 h
4.0	3 h	12½ h
5.0	2 h	8 h
6.0	83 min	5½ h
7.0	61 min	4 h
8.0	46 min	3 h
9.0	37 min	2½ h
10.0	30 min	2 h
12.0	20 min	83 min
14.0	15 min	61 min
16.0	11 min	46 min

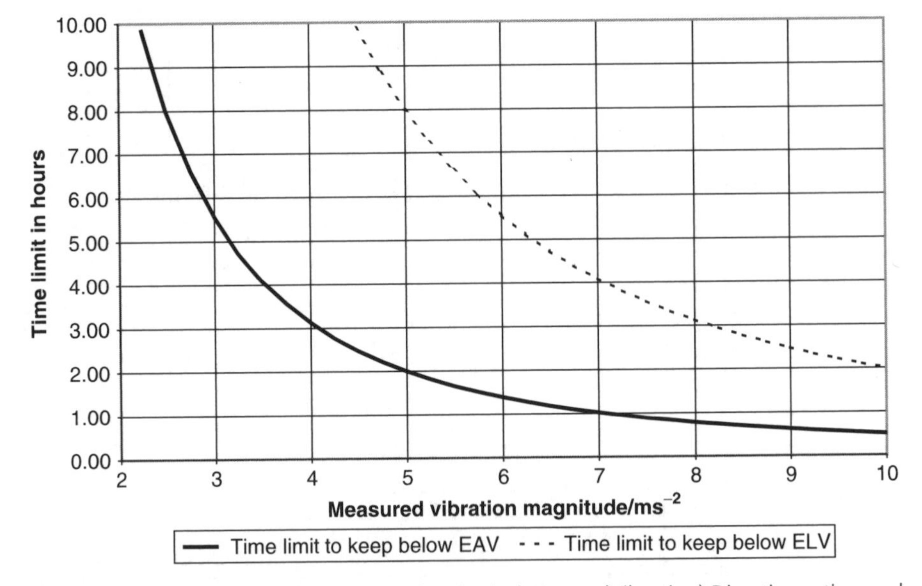

Figure 11.3 Time limits to keep below the Physical Agents (Vibration) Directive action and limit values.

12

Assessment of hand–arm vibration exposure

Is an assessment necessary?

As with noise assessments, the first step in the process of managing the risk is for an employer to recognize that some employees are likely to be exposed above the exposure action value. Most employees are clearly not at risk from hand–arm vibration and no formal assessment is necessary. The most likely ways in which an employer would become aware of a potential risk are:

- Employees who spend more than a few minutes per day using one or more of the tools which are frequently associated with HAVS;
- An actual or suspected case of HAVS among the workforce;
- Concern among employees about HAV exposure.

As with noise, the employer has a duty to arrange for an assessment of hand–arm vibration exposure to be carried out. An initial visit is likely to be necessary to establish the scope of the assessment required. This will depend on the number of tools, employees and processes involved.

It is to be expected that there will be many similarities between assessments of noise exposure and of hand–arm vibration exposure. The key difference between the two agents is that, whereas noise propagates through the atmosphere, vibration energy is transferred by direct contact. This means that:

- Only the operator will be affected by the vibration from each tool
- Vibration measurements must be made in contact with the tool handle rather than near the operator's ear
- There is no effective personal protective equipment.

Table 12.1 Some tools and processes associated with HAVS

Industry/trade	Tools
Engineering	Grinders – bench and hand-held Polishers Linishers Chipping hammers Impact wrenches Swaging
Forestry/ Horticulture/ Agriculture	Chain saws Brush saws Strimmers Lawn mowers Scarifiers Hedge trimmers
Quarrying/ Construction/ Civil engineering/ Demolition	Breakers Rock drills Stone saws Jackhammers Profile grinders Needle scalers
Miscellaneous	Leather-pounding machines Motorbikes

In turn, there are a number of consequences for the management of exposure assessments:

- Whereas noise assessments can normally be carried out as workers go about their normal duties, HAV measurements are made with the transducer in contact with the tool handle and connected via a cable to the person carrying out the measurement. This makes it much more difficult to continue as though nothing out of the ordinary was happening, and it is much more likely that simulated work operations will be specially set up so that HAV measurements can be carried out.
- The actual exposure time is frequently more difficult to determine in the case of hand–arm vibration than in the case of noise exposure assessments.

Planning a workplace hand–arm vibration exposure assessment

Before embarking on an assessment of hand–arm vibration in a particular workplace or department, information needs to be available about the size of the task involved so that the work can be planned and the time to be spent on the survey can be assessed. Much of the advice in Chapter 6 is also relevant to hand–arm vibration exposure assessments and will not be repeated here. The main difference is that only those using hand-held or hand-operated power tools will be affected by their vibration. The number of measurements to be carried out depends therefore on the numbers using power tools and the variety of the tasks involved rather than simply on the numbers working in each area.

As with a noise survey, an initial visit to determine the quantity of work involved will normally be desirable. The choice of suitable equipment can be made at this stage. Because of the physical attachment required, hand–arm vibration equipment is more likely than a sound level meter to be damaged during measurements and this may affect the choice of equipment as well as the fees to be charged by an external consultant.

The cable connecting the accelerometer to the vibration meter is particularly vulnerable, and a spare cable is essential whenever important measurements are planned. The variety of tool handles encountered means that one method of attaching the accelerometer may not work in every case and alternatives need to be available.

As with noise assessments, it is good practice to record all measurements on a survey sheet which organizes the information so that it is easy to make comparisons between different tools and individuals. Some assessments will be of the various tools used by one operator during a typical day, but it is more likely that measurements will be centred on a particular tool or operation.

Planning an assessment on a particular tool or process

The number of measurements which apparently need to be carried out for some tools can be daunting. A skilled grinder may have a large collection of wheels, discs and stones which are used for different tasks, or even for different stages of the same operation. A gardener may cut grass under a wide range of different conditions, from a bowling green on a sunny day to an overgrown verge in the rain. The factors which could potentially affect the vibration level at the tool handle could include:

- The type of tool (grinder, jackhammer, strimmer)
- The model in use (including factors such as its power and size)
- The cutting tool attached (shape of abrasive wheel, type of abrasive, grit size, tooth size, etc.)
- The sharpness of the blade
- The tool's age and state of maintenance
- The material being worked and its condition (wet, dry)
- The individual operator's technique
- The mental state of the individual operator (a tired worker may use the tool differently)
- The position and angle at which it is worked (a grinder used on the underside of a vehicle may be operated differently – even by the same operator – than the same tool used at a work bench).

Even measurements under laboratory conditions on different tools of the same model being used in the same way result in a certain spread of results. To make measurements on every possible combination of the variables listed above would take an enormous amount of time and entail unreasonable costs. It would become clear long before such a series of measurements was complete that new measurements all fell into a certain range which could have been established as a result of a much more manageable programme of work. Much of the time wasted – ultimately paid for by the employer – would have been devoted to repeated measurements on tools whose operators turned out not to be exposed to a particularly high daily equivalent. This time would have been better spent investigating the effect of these variables in cases where the additional data were needed to effect a reduction in exposure.

Having established that measurements do not need to be made under every possible combination of circumstances, it is necessary to choose which measurements to make. It should be remembered that the daily exposure will be determined mainly by:

1. Those tasks or processes which expose the operator to the highest vibration magnitudes;
2. Those tasks or processes to which the operator is exposed for the longest time.

Operators usually have a very good idea of the circumstances under which they feel most vibration, and they will certainly know which abrasives are used most frequently and for the longest periods. If this information is not forthcoming, then it may be deduced from direct observations of the work or by consulting production and maintenance records. It is normally possible to choose a representative set of conditions under which to make measurements.

Repeat measurements made using the same tool and material are useful to establish the repeatability of measurements.

If equipment is available which can make simultaneous measurements on all three axes, then the overall measurement time is potentially reduced by a factor of three. This assumes that the triaxial accelerometer is compact enough to attach to all the tool handles which could be measured with a single-axis accelerometer. Some operations are inherently of short duration and it may be difficult to make three separate measurements. In most cases this very short exposure duration is unlikely to result in a daily exposure approaching the exposure action value.

Measurement issues

The accelerometer should be mounted on the handle of the tool, or to the piece being worked where this is held by hand, as near as possible to the point of entry of vibration energy to the hand. It is important that no additional safety problem is introduced by the presence of the accelerometer, and that so far as possible it does not interfere with the normal use of the tool.

Deciding on an appropriate point and method of attachment is not always easy, as these requirements tend in practice to be contradictory. Bear in mind that there needs to be space to mount the accelerometer with three different orientations. There is a great deal of advice on accelerometer mounting in Part 2 of ISO 5349. However, the accelerometer positions shown for particular types of tool are taken from standards intended to generate data for comparing different models of tool – the best position for the accelerometer should be decided after observing the tool in use and in discussion with the operator.

Where the hand grips a tool casing or handle with a relatively large diameter, it is often possible to place the accelerometer on the side of the handle diametrically opposite the palm of the hand. In other cases it may be necessary to place it immediately adjacent to the point at which the hand grips (Figure 12.1).

Figure 12.1 An accelerometer attached at the point where the handle is gripped.

It is not unknown for operators to hold a tool in a way not intended by the manufacturer. There are frequently health and safety implications when this happens which are not related to vibration exposure. It may be that the tool has been badly chosen for the task, or that operator training is inadequate. So far as vibration exposure is concerned, though, there is no value in making measurements other than at the point where the tool is actually held (Figure 12.2).

Whatever position is chosen there are various ways in which the presence of the accelerometer might affect the safety and/or normal operation of the tool:

- Obstructing the power switch
- Locking 'on' a safety switch (e.g. some tools are supposed to cut out unless both handles are gripped)
- Preventing complete operation of a brake or clutch lever
- Limiting access to certain parts of the work due to the physical obstruction represented by the accelerometer
- Forcing an awkward orientation of the hand and fingers.

Figure 12.2 A hand-held attachment block used on an awkward shaped handle.

Figure 12.3 An accelerometer attached to a scrap component for measurements on a pedestal grinder.

In cases where the workpiece is held by hand and applied to a rotating tool, the difficulties are if anything greater. In normal use the operator may grip the work at continually changing positions, and may turn and move the workpiece continuously so that the whole surface is worked. This makes it difficult to make measurements for very long close to the point of grip, and it raises the likelihood of accidents to the operator or damage to the equipment caused by the accelerometer and its cable fouling the machinery. There is no simple set of rules for carrying out measurements in this situation. It may be necessary to make several measurements at different positions and judge afterwards which are the most representative. Alternatively it may be possible to set up an operation specially for the measurement which is as similar as possible to the live operation, but eliminates some of the possible hazards and difficulties (Figure 12.3).

For two-handed tools, attachment positions and methods will need to be chosen for both hands.

There is a standard way of orienting the three axes, which can be found in ISO 5349, and which can use either:

- biodynamic co-ordinates, based on the geometry of the hand itself; or
- basicentric co-ordinates, aligned more precisely on the tool handle (Figure 12.4).

In practice, the difference between the two co-ordinate systems is not very important. Research work must use one of these sets of axes in order for results

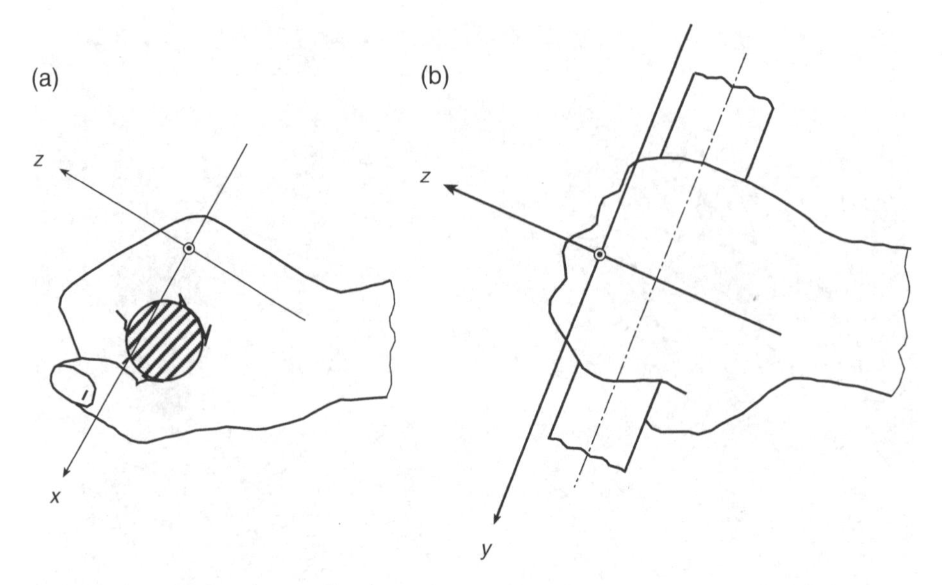

Figure 12.4 The hand–arm vibration measurement axes.

to be comparable with those of other researchers, but this is unimportant for an overall hand–arm exposure measurement, as data from the three axes are to be combined. It *is* vital that the three measurements are made along three orthogonal (i.e. at right angles to each other) axes, but their precise orientation relative to the hand is not crucial to arriving at a correct HAV measurement. However, it is good practice to make measurements if at all possible which are roughly aligned with the standard axes, as this allows easy comparison of results from similar tools, which may be important for spotting errors. In the case of the y-axis, this is normally easy as this axis is aligned along the tool handle. The positions of the x- and z-axes, though, may be transposed if the tool is gripped in a different way, and practical problems in attaching the accelerometer may mean that the actual measurement axes fall between the x- and z-axes as strictly defined (Figure 12.5).

The cable from the accelerometer must be taped to the handle of the tool to prevent inaccurate measurements – this is normally desirable, too, for safety reasons.

dc shift

'dc shift' is a problem which can occur when a piezoelectric accelerometer is used to measure vibrations with a strong impulsive content. This would include the vibrations generated by the operation of impact tools, such as hammer drills and concrete breakers. It is caused by the generation within the accelerometer of a

Figure 12.5 A tool on whose handle the x- and z-axes may become transposed.

succession of short-lived dc voltages superimposed on the ac signal. When rms averaging is applied, the dc component results in an apparently very high acceleration reading. In fact the value recorded is often so high that it is obviously incorrect.

dc shift is notoriously unpredictable but should be watched out for when making any measurements on impact tools. The solution when it occurs is to insert a mechanical filter between the accelerometer and its mounting block. This consists of a rubber block sandwiched between two sheets of metal and is carefully designed to filter out the high frequency components in the shock which generate the dc shift, without affecting the frequency range of interest. Figures 12.6 and 12.7 shows the construction of a mechanical filter. It will normally be necessary to measure along one axis at a time when using a mechanical filter, even if a triaxial accelerometer is normally used (Figure 12.8).

Other sources of measurement error

dc shift is perhaps the best known source of measurement error, and is sometimes blamed for errors which are really a result of sloppy procedure. In most cases, when the accelerometer is poorly mounted on the tool handle, the measured acceleration will be greater than the true value.

Figure 12.6 A mechanical filter.

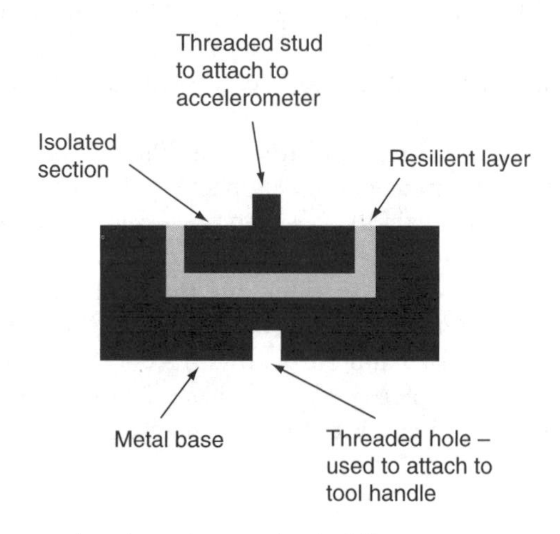

Figure 12.7 A cross-section through a mechanical filter.

A damaged accelerometer cable can also result in unrealistically high acceleration measurements – sometimes persisting when the accelerometer is not attached to a source of vibration. The obvious diagnostic test for this is to replace the cable when the meter reading should stabilize.

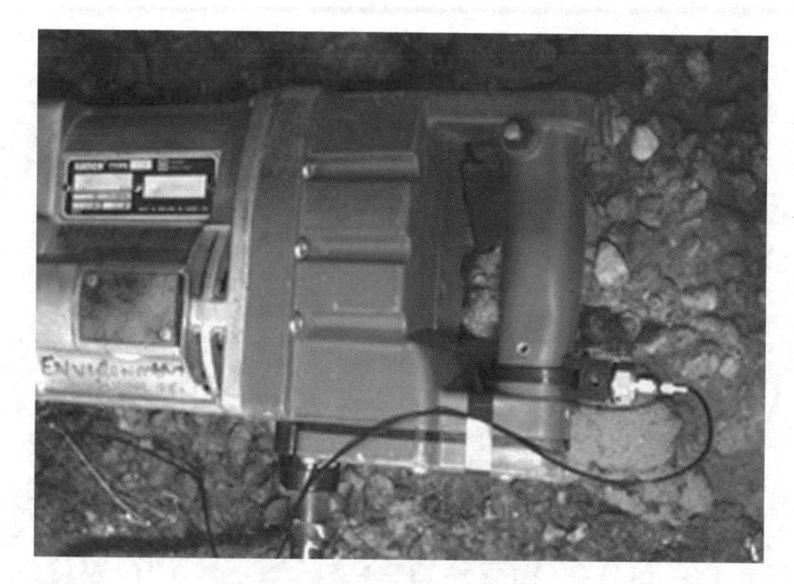

Figure 12.8 A measurement using a mechanical filter.

Mass loading occurs when the accelerometer (including any mounting block) has a mass approaching that of the tool handle it is attached to. ISO 5349 Part 2 recommends that the total mass should be less than 5 per cent of the tool handle. Above this, the mass of the handle plus block plus accelerometer will be significantly different from the handle alone. The natural vibration frequencies will change, and under certain circumstances this will lead to the measured acceleration being very different from that when no accelerometer is present.

When assessing whether the accelerometer reaches 5 per cent of the mass of the handle, it can be difficult to decide exactly what constitutes the handle without dismantling the machine. Many hand tools have an integral case and handle without internal isolation. In this case the whole tool can be treated as a unit and it is highly unlikely that the mass of the accelerometer will be 5 per cent of this. Mass loading may be important with small tools such as hand engravers, or where a lightweight handle is substantially isolated from the rest of the tool. This is the case with some horticultural tools.

Some tool handles are fitted with resilient grips which can lead to erroneous measurements. Hard pvc grips are not a problem here, but soft rubber grips when compressed by a cable tie will act like a mass-spring system with resonant frequencies in the range where they can inflate measured accelerations. Various solutions may be available, and are discussed in ISO 5349 Part 2. Essentially it is necessary either to measure on the handle in the absence of the rubber grip, or to clamp the accelerometer block very tightly – for example, using a hose clip – so as to compress the grip against the handle (Figure 12.9).

Figure 12.9 Accelerometer positioned to avoid a resilient grip.

Reading data from a hand–arm vibration meter

Most hand–arm vibration meters offer a number of quantities which can either be displayed simultaneously or from which the display can be selected. When data are stored or printed out all these possible quantities are normally included. Since the 8-h equivalent level A(8) is the quantity to be assessed, and since most meters on the market offer this quantity, it is tempting to think that this value can be taken straight from the meter. In fact, the A(8) value, along with some of the others displayed, may be calculated on the basis of some utterly unrealistic assumptions. Some hand–arm vibration meters will record an A(8) value which assumes that the measurement duration was the same as the daily duration of exposure to that vibration level. This is extremely unlikely to be correct. Others prompt the operator for an exposure time, but although this will result in a correct A(8) for that exposure time it is of little help when considering possible variations in tool use from day to day. It would be technically possible to manufacture an instrument which could take into account all possible variations and also the use of more than one tool, but entering complex data to a hand-held instrument can be awkward.

Computers, on the other hand, are designed for easy data entry. If the vibration magnitude along each axis has been measured, averaged over an appropriate time period, then all the other figures which are required can be calculated by a simple spreadsheet. It is probably best to select a measuring instrument which is optimized for making accurate measurements and to use a computer to calculate all the possible permutations of exposure.

The quantities which need to be recorded as part of a hand–arm vibration exposure assessment are the vibration magnitudes along the three measurement axes, referred to in ISO 5349 as a_{hwx}, a_{hwy} and a_{hwz}. The combined axis acceleration, a_{hv}, can be calculated from these, although it may also be helpful to have this quantity displayed. Many hand–arm vibration meters on the market were designed before this standard was published, and the display will not necessarily use ISO 5349's notation for these quantities, although the meaning should be clear if the instrument's manual is consulted. There is an additional problem with notation here. ISO 5349 does not make it fully clear whether these descriptions should be applied only to a steady vibration, or whether they refer to a quantity averaged over a specified time period (when measuring noise levels, this corresponds to the difference between the sound pressure level, L_p and the equivalent continuous level, L_{eq}).

The peak acceleration during the measurement period is often available to be displayed, and this is useful because it shows whether there is a risk of overloading the instrument during the measurement – an actual overload will be indicated as such on the display.

Measurement time

ISO 5349 Part 2 recommends that one minute should normally be the minimum measurement time. If single axis measurements are being made then this will be the minimum time per axis. Some operations are of very short duration, in which case much shorter measurements may be all that is available. More frequently, there is a regular cycle in the operation being carried out. Figure 12.10 shows an example of this. In this case the measurement time will be synchronized with the work operation. Normally the measurement duration will be two or more complete cycles, although this will depend to some extent on the time taken for each operation.

In many cases, there will be no regular pattern, and the measurement time needs to be chosen so as to average out the fluctuations in vibration. If an appropriate period cannot be decided by observing work in progress, then it may be necessary to repeat the measurement to check that consistent results are obtained.

Information collection

Along with the measurement data, information will be needed about how long the worker spends each day exposed to a particular vibration. This information can be collected by:

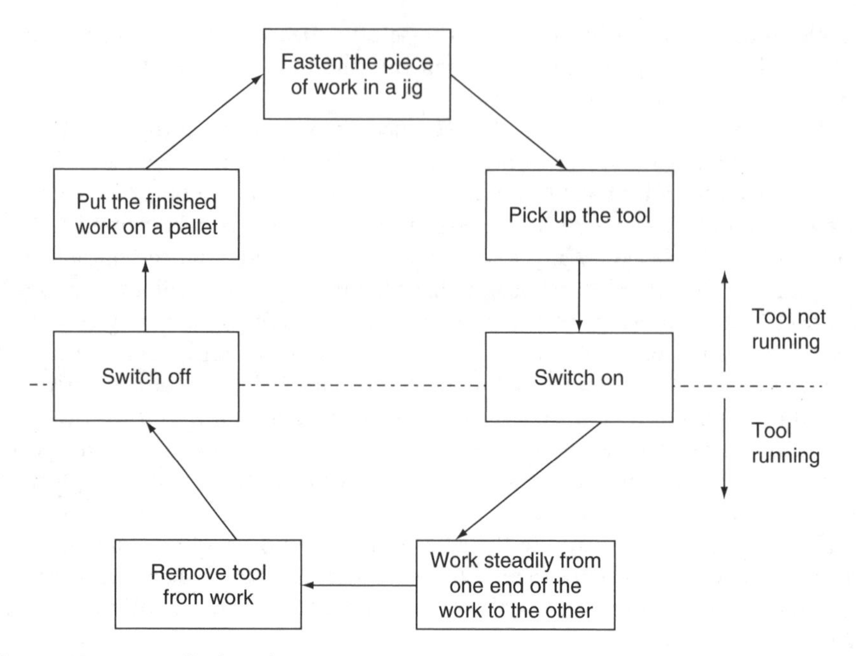

Figure 12.10 A cyclical work pattern.

- Asking the worker
- Asking the department manager
- Direct observations
- Consulting production records (in the case of some manufacturing operations) or maintenance schedules.

Clearly the most reliable conclusions will be drawn by doing all of the above and finding that they all agree. As they rarely do agree, judgements will have to be made about which information is the most reliable.

If a worker regularly uses more than one vibrating tool in the course of a shift, then information will be required about the exposure time on each. Frequently this will vary from day to day, and it may be that an increased time on one tool will reduce the time available to use others.

In many cases, at least two exposure times will emerge – one corresponding to a typical day and one which represents the maximum likely daily exposure. Recent research (Palmer et al., 1999a) shows that a direct question to the worker, such as 'How long do you spend using this tool in a typical day?', will almost certainly lead to exposure times which are overestimates, perhaps by a factor of two or three. One reason for this (but not the only reason) is that this type of question will elicit an answer based on the total time working at that task. However, using the example of Figure 12.10, the vibration exposure only occurs during the operations in the bottom half of the diagram, whereas the boxes in the top half may represent a high proportion of the total time involved. The terms 'anger time' and 'trigger time' are both used to describe the actual time spent in contact with the vibration source. The most accurate estimates of anger time probably come from direct observation of the job, followed by discussion with the operator of the number of jobs completed in a day, confirmed where possible by consulting production records.

Assessing doses and time limits

The aim of any assessment will be to calculate the existing hand–arm vibration exposure of each affected employee, and to recommend a range of possible management options. The outcome of a noise exposure assessment is frequently to indicate which action value is exceeded and to recommend the use of hearing protection. Because no effective personal protective equipment is available in the case of hand–arm vibration, it is necessary to control exposure by other means, and the most immediate solution is to limit exposure time. A series of recommended exposure time limits is therefore very often the key outcome of a hand–arm vibration survey.

Because of variability in the work load, it may not be possible to end up with a definite daily equivalent exposure A(8). Instead, it should be possible in this case to calculate a typical and a worst-case A(8). Where multiple tool use takes place, the time limit on a particular tool will depend on which tools have already been used that day and for how long. It may even depend on which tools will be unavoidably used during the remainder of the shift.

Many of those commissioning assessments of hand–arm vibration exposure would be unimpressed by a complicated array of mathematical relationships and flow charts, but neither would they welcome a set of very simple but unnecessarily restrictive time limits which would be expensive to implement. To distil a mass of measurements into a set of clear outcomes, and to further derive reliable and achievable ways of reducing it to the chosen target level is probably the key skill at this stage in the process of managing hand–arm vibration risks.

Uncertainties in assessing hand–arm vibration exposure

A hand–arm vibration assessment, like any other measurement based assessment, must include an indication of the uncertainty in the findings. The uncertainty in deriving an equivalent 8-h exposure A(8) will include contributions from:

- The declared measurement accuracy of the instrument(s) used
- The repeatability of successive measurements of the same operator using the same tool
- Differences in operator technique
- Uncertainty in exposure time

- The assumption that the process as measured is typical of the process carried out in practice (due to truncated operations for measurement convenience, interference by the accelerometer and its mount with normal control of the tool, etc.)
- Differences between apparently identical tools during their life cycles.

Some of these sources of uncertainty will be much greater than others, and in particular situations some may be much more important than is normally the case. Commonly, results can be regarded as having an uncertainty of ±30–40 per cent. In some situations, where measurement or other data collection is difficult, uncertainties may be greater than this. Under precisely controlled conditions uncertainties may be rather lower, although the practical minimum uncertainty can be taken to be ±20 per cent. A realistic assessment of the various factors leading to measurement uncertainty can only be made at the time of measurement.

Reporting hand–arm vibration assessments

Many of the recommendations of Chapter 6 can also be applied to the reporting of hand–arm vibration assessments. Some of the key points made in that chapter are here amended to apply to hand–arm vibration assessments:

1. It must clearly identify the scope of the assessment in terms of the sites, departments and individuals covered.
2. The measured levels and the exposure times estimated must be stated, although for a large or complicated assessment this detailed information is probably best put in an appendix.
3. For each employee an equivalent 8-h exposure, A(8) – or a range of A(8)s to cover different circumstances – must be stated.
4. These A(8)s should be explicitly interpreted so it is clear which employees are exposed above the EAV and which are exposed above the ELV.
5. Time limits to keep within the action or limit value – if there is a possibility of their otherwise being exceeded – should be recommended for each tool.
6. The duties of the employer with regard to each employee should be pointed out. For example, the employees who should receive information and training about vibration and its effects must be identified.

A range of appropriate measures to reduce HAV risk will be recommended for consideration without pre-empting the manager's duty to decide which steps are reasonably practicable.

Case study 12.1 A hand–arm vibration survey

Lanshaw Engineering Ltd has 20 shop-floor employees and produces equipment for the catering trade. Following a case of HAVS among their production workers – and even though this was established to be due to HAV exposure in a previous job – it was decided to commission a survey of hand–arm vibration exposure throughout the production area.

Preliminary survey

The health and safety manager walked round each department with the relevant unit manager, identifying tools that are commonly associated with significant levels of HAV exposure. The list included hand-held and bench grinders, linishers, polishers and impact wrenches. A list was typed up and circulated to the managers for correction. Several more tools surfaced at this point and were added to the list.

An attempt was made to compile data on the HAV exposure to be expected from each machine from the manufacturers' literature. The data acquired in this way was incomplete, and investigations also revealed that this kind of information may not be a reliable guide to employee exposure.

At this point it was decided that an external consultant should be commissioned to assess the HAV exposure from each tool. This would involve measurements at each tool handle while it was being used to carry out the normal job or jobs for which it was used. Afterwards each user was interviewed about the times for which the tool was likely to be in use during a working day, about the different abrasive wheels used, and about maintenance practices.

The measurement process

Several 100-mm angle grinders of the same model were in use to carry out similar work. Measurements were confined to two of these grinders to save time. Other users, though, were asked if they had noticed high vibration levels from their grinders. None had.

In the case of the hand tools it was relatively easy to attach an accelerometer to the handle close to the point at which it was gripped. Some tools had two handles. In this case, measurements were made on each handle in turn. The higher of the two vibration magnitudes was then used to assess the daily dose and to set time limits if this was appropriate. Some tools were gripped by both hands close together around the body of the tool. In this case, the accelerometer was attached between the two hands and the measured values were assumed to apply to both.

Two tools not on the list provided (in either its original or amended version) – and of which the unit managers were unaware – were produced by employees during the measurement process. Measurements were made of the vibration levels at the handles of both these tools. In one case – a 175-mm grinder – levels were considerably higher than were measured on other grinders. Although its user thought that this grinder was particularly effective, other workers thought the same job could equally well be done with the smaller grinders provided, and it was therefore disposed of.

The factory produces machines in small batches rather than on a production line. Much of the power tool use forms part of various finishing processes, and as the quantity of work required varies with the quality of the components supplied. As a result, there is considerable variation in the time spent using any particular tool in the course of any one shift. In addition, an individual worker on some days was likely to use two or three different power tools for varying periods.

With the measurement results available, it was possible to work out the maximum HAV exposure that might occur. This could normally be done by measuring the time spent working each component, and then asking about how many times this operation would be repeated in the course of a typical day's work, and also the maximum number of pieces that could be worked in a day. The production methods in this factory meant that these numbers could later be checked against production records. A large batch of one component would mean that the HAV exposure from that tool would be maximized, but it also left less time for using other tools in the same shift. Juggling around the various exposure times led to an estimated typical daily exposure and also to the maximum daily HAV dose to which that individual would be likely to be exposed. For example, one operator was responsible for grinding out imperfections in a mixing paddle. He reported that these components normally came in batches of no more than 10. Production records, though, showed that a batch of 20 had been processed on one occasion in the previous year. The average time taken for each paddle was 20 min, so it would be possible to spend the whole of a 12-h shift on that task. If so, though, no other vibrating tools would be used that day. A batch of the more normal size of 10, though, would be completed in 5 h and it was estimated that after allowing for record keeping and setting up the machine for a new job, up to 90 min might be spent in the same shift on other work involving vibration exposure. The maximum daily exposure could therefore be assessed (a) when processing normal-sized batches; and (b) when working on a large order for that particular component.

Some components are polished on a bench-mounted polishing machine. The polishing mop in this case rotates, while the operator holds the component in gloved hands, turning and moving it in contact with the mop until satisfied with the overall finish. To complete the survey, a measurement of the HAV exposure involved in this operation was required, but it was clearly a difficult measurement to make. A number of reject components was found, and an aluminium block was glued to one of them. The operator was asked to polish one side of the component only, but to apply it in other respects as if he was polishing a real component – this procedure was practised without the accelerometer attached. Finally, the triaxial accelerometer and vibration meter in use were replaced by a hand–arm vibration meter that used a single axis accelerometer. Although this increased the measurement time considerably, the single axis accelerometer is much cheaper and it was decided not to risk damage to the more expensive equipment. Because of the artificial aspects of this particular measurement, allowance has to be made for a greater degree of uncertainty in the final assessment than is normally the case.

The next stage

Once the exposure levels had been assessed, a meeting was held to discuss the outcomes. It was clear that the hand–arm vibration exposure of some employees exceeded current HSE guidelines and also the Physical Agents (Vibration) Directive limit values. Some managers were shocked to discover that no effective personal protective equipment was available. The health and safety manager argued that the substitution or elimination of processes involving high levels of hand–arm vibration exposure was the best approach, but the production manager pointed out that this could take months or years to achieve. The same manager was unenthusiastic about rotating between two or more employees those jobs entailing high levels of hand–arm vibration exposure, but was eventually persuaded that this would be the best short-term solution, while a more effective long-term strategy was developed.

Case study 12.2 dc shift

A worker in the packing department of Doubler Building Products Ltd spends a large part of the day assembling wooden pallets. The premachined components are first placed in a jig, and then a nail gun is used to fasten them together. The nail gun is therefore in operation for a period of about 30 s, repeated every 2 min, while pallet assembly is in progress.

A hand–arm vibration meter was used to assess the vibration exposure during this operation. After the first measurement the results were checked and found to be as follows:

x-axis, $124\,ms^{-2}$
y-axis, $2.54\,ms^{-2}$
z-axis, $162\,ms^{-2}$

These values are unlikely to be correct. The values for the x- and z-axes are extremely high. The operator would experience extreme discomfort if they were correct, and would quickly suffer from hand and arm injuries in the very unlikely event that he or she were prepared to continue using the tool at all. It is also very unusual for the values for the different axes to differ by as much as is shown – the z-axis magnitude is more than 60 times greater than the y-axis magnitude – and this is another reason for suspicion.

The operator was asked about the levels of vibration and shock felt when using the tool and confirmed that they were not unusually high. A nail gun is a percussive tool – it relies on impact to drive the nail in – and so dc shift was thought to be a likely explanation. The measurements were repeated using a mechanical filter between the accelerometer and its mounting block, with the following results:

x-axis, 1.46 ms^{-2}
y-axis, 2.69 ms^{-2}
z-axis, 4.41 ms^{-2}

These new values are more credible in terms of the absolute values of the x- and z-axis magnitudes, and also in terms of the relationship between the different axes. The y-axis value is very similar to the original measurement; no dc shift was evident on this axis in the first set of measurements, presumably because the shocks when the nail gun operates are not significantly transmitted on this axis. The diagnosis of dc shift affecting the original measurement was confirmed, and all that remained was to repeat the measurement over a few more assembly cycles to arrive at a representative value of a_{hv} to use in assessing the operators HAV exposure and in deciding whether action was required to reduce it.

Whole Body Vibration

Controls on whole body vibration

What is whole body vibration?

Whole body vibration (WBV) is the name given to vibration which enters the body via a number of routes and can potentially affect organs which are not adjacent to the point of entry to the body.

The most important entry points are:

1. The feet
2. The buttocks
3. The back
4. The back of the head.

In a workplace context it is most normal for an affected individual to be either standing or sitting. When the subject is standing, the vibration energy will obviously enter via the feet. When seated, it will enter mainly via the buttocks and the feet, although in some situations it can enter via a backrest or headrest. There are situations in which a recumbent person can be affected, for example in ships, offshore platforms or long-haul aircraft which carry sleeping accommodation. In this case it would be normal to assume that the vibration energy was entering via the back.

Chapter 7, which is about how vibration is measured and described, is equally relevant to both whole body and hand–arm vibration.

Health effects

The effects of vibration exposure on the human body have been studied over a number of years. Both animal and human studies have established that exposure to high levels of vibration can have serious effects on the human body, causing damage to a variety of vital organs. However, the levels of vibration involved in this type of damage are high enough to cause a great deal of discomfort, and it is unlikely that human beings would allow themselves to be exposed to this level of vibration for extended periods. Workplace exposure to whole body vibration involves vibration magnitudes and exposure durations which may involve mild discomfort but which are also suspected of having rather more subtle long-term effects on health.

Whole body vibration has been linked with the following effects on human beings:

- Perception
- Discomfort
- Interference with vision
- Interference with fine motor tasks
- Spinal injuries
- Damage to the digestive system
- Damage to the reproductive system.

Some of these – perception, for example – cannot really be described as health effects. Others are controversial. The effects on the digestive system, for example, have not been demonstrated beyond doubt. Whole body vibration exposure which falls short of the magnitudes and durations which can definitely lead to health effects may nevertheless cause annoyance, reduced work efficiency and loss of concentration.

When dealing with noise it is normal to separate the possibility of direct health effects (i.e. hearing damage) from the other effects of noise. Legislators have seen it as their role to protect employees' hearing, while leaving it to the employer's good sense and commercial pressures to ensure a contented, productive workforce. In the case of whole body vibration, though, there is a much smaller difference in amplitude between a vibration level of which the recipient is just aware, and a vibration level which causes serious annoyance. It is also rare that vibration serves a useful purpose, while noise is an unavoidable and frequently useful part of our everyday environment. Vibration which causes annoyance, particularly if it originates outside an employer's premises, can be compared with the various sources of environmental noise. These, too, can cause annoyance and as a result may affect health, but they are unlikely to lead directly to physical damage. Noise and vibration can affect individuals at home or in leisure pursuits, as well as at work, and the effects of unwanted vibration on an office worker, for example, can be compared with those on someone watching television at home. Both may be justifiably annoyed by the vibration, but we would not expect the health of either to be adversely affected.

ISO 2631 and its dose–effect relationships

The current international standard on the assessment of whole body vibration is ISO 2631: 1997. Part 1 deals with direct health effects, while Part 2 deals with vibration transmitted through buildings and is more concerned with nuisance vibrations. Part 4 deals with the effects of vibration on train passengers and crews. Like most standards, ISO 2631 has been developed over a number of years and through a number of different draft standards. Even so, it has been criticized on the grounds that it contains material which is not really justified by the current state of scientific knowledge of the health effects of whole body vibration. The British Standards Institute, for example, has not so far adopted the International Standard and the current UK equivalent is BS 6841:1987. In practice, and particularly now that a European directive is in place, it is not realistic to ignore the existence of an international standard and even in the UK ISO 2631 is probably used more widely than BS 6841.

The thresholds of perception and discomfort are, like all vibration effects, likely to vary a great deal between less- and more-sensitive individuals, and will also depend in a single individual on factors such as state of alertness, presence or absence of distracting factors and on emotional state. ISO 2631 aims to establish levels at which typical individuals will report that they can feel a vibration, that they find it uncomfortable, or that adverse health effects can be predicted.

Interference with fine motor tasks is best illustrated by attempting to drink a cup of coffee on a train. More seriously, in jobs where fine manipulation is required it can be dangerous for a vibration-exposed machine operator to be unable to use controls accurately. The ability to read information accurately from display devices is a similar concern when considering the effects on vision.

Effects on internal organs, such as the digestive and reproductive systems, have been reported, but the link is not well established. Some researchers have concluded that vibration levels found in workplaces are unlikely to have such effects. ISO 2631 does not attempt to predict the likelihood of these effects if they exist.

Probably the most important of the health effects addressed by the standard is the link with spinal injury. This, too, is controversial. Although it is not normally denied that this sort of injury can be caused by WBV exposure, the relative importance of WBV and other agents has not clearly been determined. Most of those exposed to WBV in their jobs are in manual trades which expose them to other potential hazards. Heavy lifting, and prolonged periods sitting in poorly designed seats are also common causes of back injury. In some cases vibration exposure and poor posture go together. Tractor drivers, for example, may be exposed to relatively high WBV, but they also frequently have unsuitable seats, and some of the tasks they perform result in poor posture.

Whole body vibration exposure in the workplace is normally associated with driving, or being carried in, vehicles. Off-road vehicles used in agriculture and construction are particularly likely to cause significant exposure, although there is a great deal of variation even between machines that do the same job. Apart from vehicles, some machinery used in quarrying and mineral extraction, and in the concrete industry may involve workers in similar exposures. Helicopter crews are similarly exposed to high vibration levels, though occupants of fixed-wing aircraft generally are not.

Although an assessment of an individual's whole body vibration exposure may contribute to an ergonomic assessment of a job, it would be unwise to use a WBV assessment on its own to predict the likelihood of back injuries or to develop a programme for preventing them.

The assessment of whole body vibration is inherently a complicated matter, and this is reflected in the complexity of the standard. Three axes are measured, based on the orientation of the human body (Figure 13.1):

- The x-axis is in the back-to-front direction
- The y-axis is from side-to-side
- The z-axis is from feet to head.

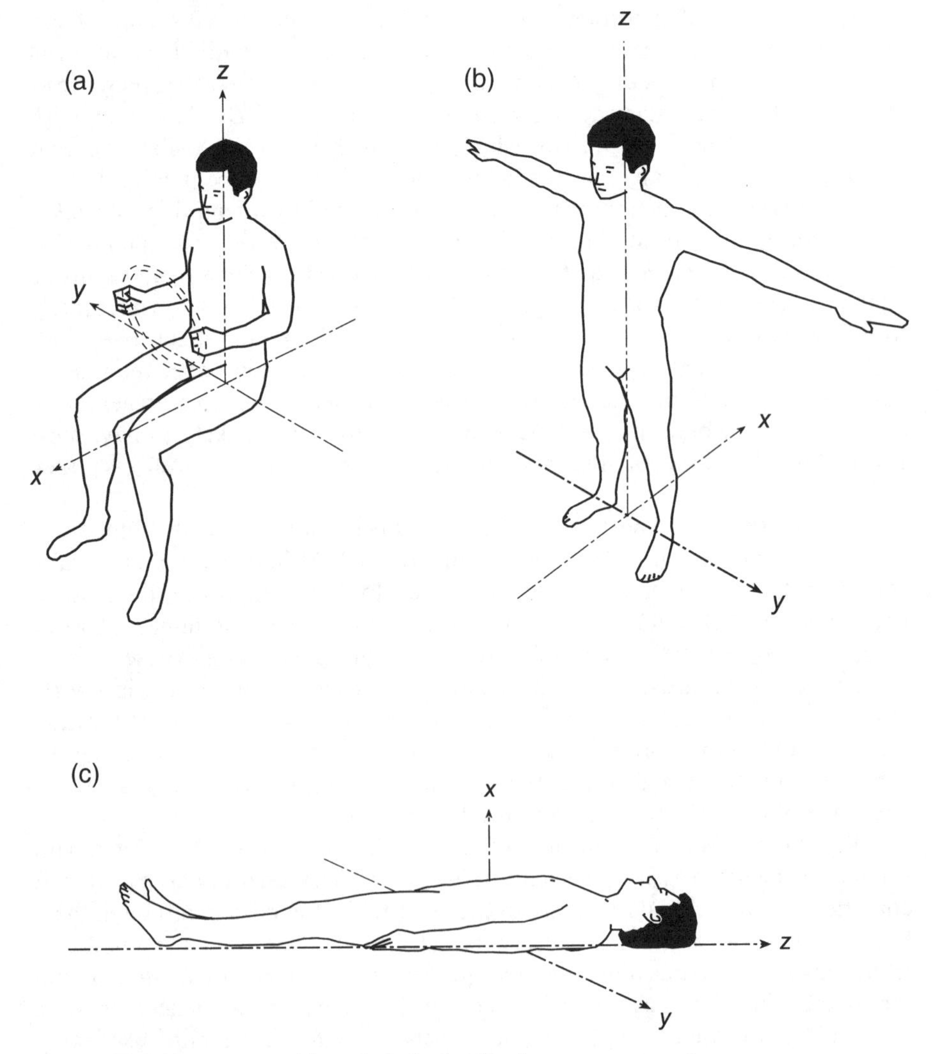

Figure 13.1 The axes used for whole body vibration measurement.

For a recumbent subject, the above orientations are maintained, so that although the z-axis is normally the vertical one, in the case of a recumbent subject it becomes one of the horizontal axes. The next step when assessing hand–arm vibration exposure was to combine data from the three axes so as to arrive at a combined axis assessment. Whole body vibration assessments do not normally do this, but instead assess each axis separately.

ISO 2631 defines no fewer than six frequency weightings to be used in different circumstances. They are designated W_c, W_d, W_e, W_f, W_j and W_k. Fortunately, most assessments can be carried out using just two of them: W_d for the x- and y-axes and W_k for the z-axis (Figure 13.2). W_f is only used for assessing the likelihood of vibration exposure causing travel sickness, and it is defined over a range of lower frequencies than the other weightings.

Along with the various frequency weightings, ISO 2631 specifies a number of multiplying factors to be applied to the vibration magnitude once it has been measured using the appropriate frequency weighting (Table 13.1). For workplace exposure assessments, concentrating on possible health effects, values measured along the horizontal (x and y) axes must be multiplied by a factor of 1.4. Measurements on the vertical (z) axis are multiplied by a factor of 1 (in other words, in practical terms, they do not need to be multiplied by anything).

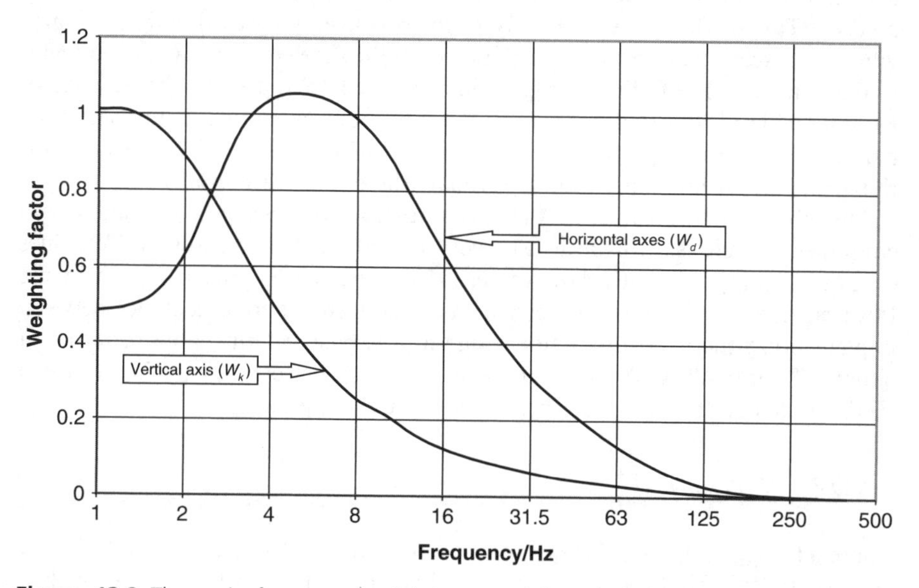

Figure 13.2 The main frequency weightings used for whole body vibration measurement.

Table 13.1 Frequency weightings and scaling factors for whole body vibration assessments

Axis	Vibration effect to be assessed	Point of entry to the body	Frequency weighting	Multiplication factor
x	Health	Buttocks	W_d	1.4
y			W_d	1.4
z			W_k	1
x	Comfort	Buttocks	W_d	1
y			W_d	1
z			W_k	1
All	Comfort	Any other point	Consult ISO 2631 for detailed requirements	
z	Motion sickness	Feet, buttocks	W_f	1

The type of time averaging which it is most appropriate to use when assessing WBV exposure has been the subject of considerable research. The simplest approach is to use the equal energy principle, using an rms average should be evaluated over a standard 8-h assessment period. The quantity which is used to assess daily exposure using this technique is the equivalent 8-h continuous exposure. As with hand–arm vibration assessments, it is given the abbreviation A(8). The equivalent 8-h exposure can readily be calculated from measurement data covering shorter, but representative, portions of the day.

However, research has shown that equal energy averaging as described in the previous paragraph underestimates the effect of high vibration magnitudes as compared to extended periods of exposure. To take account fully of these episodes of high vibration magnitude, root-mean-quad or rmq averaging can be used. This works in a similar way to the rms averaging used with hand–arm vibration, except that now the fourth power of the acceleration is averaged and it is the fourth root of this average which is used to represent the equivalent continuous level. Put another way, if the vibration magnitude is doubled, then a quartering of the exposure period (as is used with noise and hand–arm vibration exposure) would tend to underestimate the human response.

The vibration dose value, or VDV, is a quantity which can be measured with equipment which uses this fourth-power integration. Numerically, VDV values cannot be compared with A(8) values. However, the European Union in the Physical Agents (Vibration) Directive has given member states a choice between implementing the exposure action value and exposure limit value in terms of either A(8) or of VDV. A VDV for an entire day's work can be calculated from a VDV measured during a shorter, but representative, period.

$$\mathrm{VDV} = \mathrm{VDV}_{\mathrm{part}} \times \sqrt[4]{\frac{T}{t}} \qquad (13.1)$$

where $\mathrm{VDV}_{\mathrm{part}}$ is the VDV measured over a representative period t and T is the length of the full shift (not a standard 8-h shift). The unit in which the vibration dose value is measured is 'metres seconds to the minus 1.75', written as $\mathrm{ms}^{-1.75}$ or $\mathrm{m/s}^{1.75}$.

If the measuring equipment uses rms averaging then a true VDV cannot be calculated from the measurement results. In some cases, though, it is possible to calculate a quantity known as the estimated VDV or eVDV from rms measurement data.

$$\text{eVDV} = 1.4 \times a_{rms} \times \sqrt[4]{t}\ \text{ms}^{-1.75} \tag{13.2}$$

where a_{rms} is the acceleration measured over a representative period with equipment which uses rms averaging; and t is the length of the shift, measured in seconds. The eVDV can be close to the true VDV value as long as the crest factor of the vibration is small – below about 6 (in other words as long as significant shocks and jolts are not present in the vibration). It is normally easier to measure than the true VDV as it does not require the use of equipment which can apply rmq averaging.

Example

Whole body vibration measuring equipment is fitted to the seat of a fork-lift truck for a period of 1 h. The recorded VDV in that time is 6.2 ms$^{-1.75}$. The rms averaged acceleration is 0.52 ms^{-2}. Both values are recorded for the z-axis. Predict the VDV to be accumulated during an 8-h shift spent driving the same vehicle, and compare this with the eVDV value predicted from the rms measurement.

$$\text{VDV} = \text{VDV}_{\text{part}} \times \sqrt[4]{\frac{T}{t}} = 6.2 \times \sqrt[4]{\frac{8}{1}} = 10.4\,\text{ms}^{-1.75}$$

$$\text{eVDV} = 1.4 \times a_{rms} \times \sqrt[4]{t} = 1.4 \times 0.52 \times \sqrt[4]{8 \times 60 \times 60} = 9.4\,\text{ms}^{-1.75}$$

Because the exposure continues for 8 h, the equivalent 8-h level in this case would be 0.52 ms^{-2}. Both the VDV and the equivalent 8-h level are slightly higher than their corresponding exposure action values.

The Physical Agents (Vibration) Directive

The 2002 Physical Agents (Vibration) Directive establishes an exposure action value and an exposure limit value. For whole body vibration, each quantity can be evaluated either in terms of an equivalent continuous 8-h exposure or in terms of a vibration dose value.

- The exposure action value is an equivalent 8-h exposure of 0.5 ms^{-2}, or a vibration dose value of 9.1 ms$^{-1.75}$.
- The exposure limit value is an equivalent 8-h continuous exposure of 1.15 ms^{-2}, or a vibration dose value of 21 ms$^{-1.75}$.

Member states are free to use either the VDV values or the equivalent 8-h values when framing their domestic legislation. At the time of writing (early in 2004) this choice was still being made. Early indications are that most states will choose the 8-h equivalent method. In the UK, the Health and Safety Commission is proposing as part of a consultation exercise that the exposure action value should be expressed in terms of the VDV, but that the 8-h equivalent should be used for the exposure limit value (HSC, 2003). The duties on employers at these levels are the same as for hand–arm vibration, and are listed in Table 10.2, as well as in Appendix B.

The precise values chosen for the exposure action and limit values resulted from an extended period of lobbying and negotiation between the interested parties. They have been criticized for being too low, and there is a body of opinion which contends that action to prevent the health effects of whole body vibration exposure is justified, but that there is not yet sufficient scientific evidence to support the establishment of precise exposure limits. Some of the critics have pointed out that vibration levels in private cars and public transport frequently exceed those set as the action and even the limit value. This particular criticism depends on confusing a short-term vibration exposure with an equivalent 8-h exposure. There is in any case evidence of health effects on those who spend long periods driving cars, although these effects may not be wholly or even mainly caused by the vibration present.

Partly as a result of the lobbying by those with an interest in the outcome, the European Parliament agreed an extended implementation period for the whole body provisions of the directive:

- By the summer of 2005, all member states must have passed domestic legislation giving employees protection against vibration risks which is at least as stringent as that in the directive.
- In most industries, WBV exposure resulting from machinery purchased after the summer of 2007 is governed by the directive's provisions from the summer of 2008.
- In the case of the agriculture and forestry industries, WBV exposure resulting from equipment purchased up to the summer of 2007 is covered from the summer of 2011 onwards.

Measurement and assessment of WBV exposure

Assessment and measurement

Under the Physical Agents (Vibration) Directive, it will be a duty of employers to carry out an assessment of the risks to any employee whose work is likely to expose him or her to whole body vibration. The first step is to decide which employees are so exposed. This would include:

1. Any employee whose job involves extended periods driving vehicles
2. Any employee whose job involves driving vehicles off-road
3. Any employee engaged in static work where significant levels of whole-body vibration are apparent.

The third group above is relatively small. It includes numbers of workers operating various machines in the quarrying, mining, aggregate handling and concrete industries. It would not, for example, include workers in a workshop in which a low level of vibration was transmitted by machinery into a solid floor.

Having identified the affected employees, it has to be decided whether measurements are necessary. When assessing exposure to noise or to hand–arm vibration measurements are almost always required. However, much exposure to whole-body vibration takes place under relatively controlled conditions, including the driving of road vehicles on public highways, and of many other types of vehicles on level, hard surfaces.

The first step in an assessment should be the collection of data on whole body vibration exposure on that particular vehicle. This may be available from the supplier of the vehicle, or from a published database. As far as possible it should relate to a realistic pattern of use for that vehicle. Failing this, it will be necessary to use test track measurement results.

In many cases, after taking into account the extra uncertainties involved in using published data to real work situations, the data collected will be sufficient to arrive at an assessment of exposure. In other cases, and particularly when dealing with off-road vehicles or with older vehicles for which test data cannot be obtained, it may be necessary to carry out some measurements to complete the assessment.

Equipment

Accelerometers intended to measure vibration entering via the seat are normally built into a rubber pad which fits between the seat itself and the subject's buttocks. The pad contains three mutually perpendicular accelerometers, and the whole assembly is technically called a triaxial seat accelerometer, but is known colloquially as a whoopee cushion. Although it is not always essential to measure the three channels simultaneously, the three accelerometers are normally present and it is very convenient to use a measuring and recording system which is capable of handling at least three measurement channels if this is available. Measurements need to be longer than is normal for hand–arm vibration assessments and the inconvenience of making three separate measurements is correspondingly greater. Each of the measurement channels must be capable of being calibrated to take account of slight differences between the sensitivities of the three accelerometers which make up the triaxial assembly. It must also be possible to apply the correct frequency weighting independently to each channel. For a seated subject this is straightforward as the vertical axis will almost always be monitored by the accelerometer oriented at right angles to the plane of the rubber pad (Figure 14.1). The x and y axes may be interchanged for convenient mounting of the accelerometer assembly, but the same frequency weighting is normally used for both these axes so this should not be a problem.

When backrest measurements are made, the same accelerometer assembly is used, but in this case the axes may be changed (Figure 14.2). On a seat, the direction perpendicular to the pad itself is the z-axis. On a backrest, this becomes the x-axis. The set-up of the measuring equipment will need to be changed to use the correct frequency weightings for the new orientation. Normally it is only the x-axis (perpendicular to the backrest and to the subject's back itself) which is important for backrest measurements.

For a standing subject, or to measure vibration levels at the foot in the case of a sitting subject, the best option would be to drill and tap a hole in the floor adjacent to the foot position to attach a standard accelerometer – either a single one or a triaxial assembly designed to be attached to a rigid surface (i.e. not the seat pad). Very frequently it is not possible to drill a vehicle floor, and if this is so then other mounting systems must be considered. These include:

Figure 14.1 A seat accelerometer.

Figure 14.2 A seat accelerometer used for backrest measurements.

- Attaching a threaded stud to the floor using cyanoacrylate adhesive
- Attaching the accelerometer to the floor by means of a small quantity of softened beeswax
- Using a magnet (designed for the purpose) in the case of a steel floor
- Mounting the accelerometer to one end of a metal bar, the other end of which is inserted between the floor and the operator's foot.

It is important to get as rigid a connection between the accelerometer and the floor as can be achieved. This may be difficult in cases where a carpet or resilient floor covering is present. The possibility of damage to the accelerometer itself is another important consideration when it is mounted on a floor. This may tip the balance in favour of single-axis measurements in this situation. For vibration transmitted through a control pedal, the accelerometer will need to be attached to the pedal itself, but without compromising the operator's ability to control the vehicle.

In the case of a recumbent subject, the normal orientation of the three axes will again change. The z-axis runs from the feet to the head and will now be one of the horizontal axes. The x-axis is likely to be the vertical axis. The frequency weightings applied by the measuring instrument may need to be changed to reflect this.

Sometimes the measuring instrument will be a frequency analyser which is capable of measuring and storing the acceleration in several frequency bands simultaneously. In this case, decisions about the correct frequency weighting to use can be made when the results are analysed as the frequency weighting can be applied to the stored spectrum. However, this may mean that weighted measurement results will not be available at the time of the measurement, in which case time can be wasted due to the inability to assess the value of each set of results before moving on to the next measurement.

Ideally, the instrument should be capable of applying either root-mean-square or root-mean-quad averaging (or even both simultaneously). In practice, there are few instruments on the market which are capable of root-mean-quad averaging. Some computer-based data acquisition systems are very flexible and will store data to which the appropriate time averaging can be applied later, but in many cases it will be necessary to settle for rms averaging, and checks will need to be made that this is acceptable for the vibration being monitored.

In a few cases it may be possible for the person carrying out the assessment to sit next to the subject of the measurement to control and monitor the progress of the measurement. More often it will be necessary for the equipment to be secured in place and then controlled either remotely or by setting up a sequence of measurement and data storage operations to continue automatically.

Measurement issues

As with other workplace measurements, measurements need to be long enough to be sure that a sufficiently representative sample of the individual's daily exposure has been collected. Information about the normal and exceptional work cycles can be collected by interviewing the subject, and department managers, and by direct observation.

Most significant whole-body vibration exposure results from driving vehicles, and particularly vehicles which operate off the road. There are obviously issues of repeatability here. Many factors can affect the vibration levels experienced by the operator. In practice, some of these variables are more important than others; the design of the machine is frequently more important than the terrain over which it is operated, and measurements are normally completed in considerably less than a full day's work. The equipment used is delicate and expensive, and conditions in and around this sort of vehicle are often unfavourable, and this is a reason not to prolong measurements unduly. Seat vibration measurements can be inflated if the driver gets in and out of the seat frequently, as is the case, for example, with some delivery operations. The accelerometer will detect a disturbance which may significantly increase the time-averaged reading, and yet this clearly does not arise from vibration transmitted to the driver, but from the driver's own movements. Careful timing and control of measurements will be required if this seems likely to be a problem.

The measuring chain must be able to handle the highest shock level which is experienced during the measurement. In some cases this can be much higher than the long-term average (in other words the vibration signal has a high crest factor), and this may not be obvious at the start of the measurement. Although some instruments may claim to have an auto-range capability, it is not appropriate to use this since it will normally determine the measuring range at the start of a measurement – probably before the vehicle is moving. Range changes during a measurement will normally reset the current measurement, losing previous measurement information. The way to avoid overload problems is either to carry out a series of short measurements, or at least to start with a short pilot measurement to determine the correct measurement range.

Analysing measurement data

It is not normal to combine the data from three axes, even though a procedure exists (see ISO 8041:1993) for calculating a combined value. The Physical Agents (Vibration) Directive adopts the normal practice of making a separate assessment of the vibration component along each of the measured axes. The values measured along the x- and y-axes must be multiplied by 1.4. Some triaxial vibration meters may allow this factor to be programmed in when setting the instrument up for the measurement, in which case it is obviously not necessary to make this adjustment after measurement. Some measurements of whole body vibration exposure may be carried out over a complete working shift. In other cases this is unnecessary, either because the vibration exposure occupies a relatively short part of the shift, or because vibration conditions are uniform over an extended period which can be sampled in a much shorter time.

Although some instruments may be capable of making simultaneous measurements using rms and also rmq averaging, it is more likely that only one of these quantities will be available. It is desirable that, as well as an rms vibration measurement for each axis, the peak acceleration during the measurement (or the crest factor, from which the peak value can be calculated if the rms value is known) should be recorded if possible. The availability of all this information for each measurement period should be taken into account when choosing a measuring instrument, since the longer duration of each measurement (as compared, say, with a hand–arm vibration measurement) makes it worth investing more in equipment in order to save on the user's time.

In ISO 2631:1997, an assessment based on the rms vibration magnitude is called the 'Basic evaluation method'. The standard recommends that this method will be satisfactory as long as the crest factor is no greater than 9. One approach recommended, if the crest factor is greater than 9, is to measure the rms acceleration in contiguous 1-s periods. The highest of these 1-s values is called the maximum transient vibration value or MTVV. If the MTVV is more than 1.5 times the continuous rms value, then the MTVV should be quoted as well as the continuous rms value. An alternative way of dealing with high crest factors, according to the same standard, is to measure the VDV.

The Physical Agents (Vibration) Directive does not make use of the MTVV approach, and the regulations made by each member state to implement the directive will make it clear whether rms averaging, or VDV, or some combination, is to be used. The crest factor is still a useful quantity when making rms measurements, and particularly when these are to be used to calculate an estimated vibration dose value or eVDV.

Calculation of 8-h equivalent levels, vibration dose values and estimated vibration dose values

The Physical Agents (Vibration) Directive

The Physical Agents (Vibration) Directive requires an assessment to be made of employees' exposure to whole body vibration. When the assessed value has been compared to the exposure action and limit values, the employer's duties can be determined in relation to each group of exposed employees. The duties for whole body vibration are the same as for hand–arm vibration. They are summarized in Table 10.2 and in Appendix B. For whole body vibration, either the A(8) value or the vibration dose value for each axis, after multiplication by 1.4 in the case of the horizontal axes, is compared with the exposure action and limit values. The relevant employer duties apply whether the action (or limit) value is exceeded for just one axis, or for all three. For both noise and hand–arm vibration assessments it is normal to make measurements during actual or simulated work operations, although in each case there may be circumstances where it is appropriate to use data from other sources. Whole body vibration assessments will be based in a much greater proportion of cases on published data for a particular type of machine. This may be provided by the manufacturer, or it may be produced as a result of independent tests. Caution needs to be exercised when using whole body vibration data which have not been produced for the purpose of making exposure assessments, and this issue is discussed further in Chapter 15.

It is not normally possible to take a figure for the daily exposure of an employee to whole body vibration from published data. The figures will normally be either for:

- The rms vibration magnitude averaged over a suitable period, or
- The vibration dose value measured over a specified period using rmq averaging.

This is also the information that must be generated if direct measurements are made.

- From rms data it is possible to calculate either the 8-h equivalent level A(8), or the estimated vibration dose value, eVDV.
- From the vibration dose value over a specified period it will be possible to calculate the VDV over the actual period of exposure.

In either case, it must be assumed that the measurement was made over a sufficiently long period while the vehicle was being used in a manner representative of the use to which the exposure assessment relates. For example on-road and off-road values for most vehicles will be very different.

Calculation of A(8) from rms measurements

To calculate the 8-h level equivalent to a single period of exposure to one steady level:

$$A(8) = a \times \sqrt{\frac{t}{8}}\ \mathrm{ms^{-2}} \tag{14.1}$$

where A(8) is the 8-h equivalent level; a is the measured vibration level; and t is the exposure duration in hours.

Although it is assumed here that the vibration exposure is at a steady level, this calculation can equally be used when the acceleration level is actually a time average over a representative period generated by the measuring equipment.

A more complicated version of equation 14.1 can be used if two different levels are involved in the course of a day's whole body vibration exposure (e.g. two vehicles are operated for different periods, or one vehicle is used under different conditions).

$$A(8) = \sqrt{\frac{a_{a1}^2 \times t_1 + a_{a2}^2 \times t_2 + a_{a3}^2 \times t_3 \ldots}{8}}\ \mathrm{ms}^{-2} \qquad (14.2)$$

here, a_1, a_2, etc., are the individually measured vibration magnitudes on that axis, and t_1, t_2, etc., are the corresponding exposure times in hours.

Calculation of eVDV from rms measurements

The estimated vibration dose value can be calculated in a similar manner from the rms data. For a single exposure period:

$$\mathrm{eVDV} = 1.4 \times a \times \sqrt[4]{t}\ \mathrm{ms}^{-1.75} \qquad (14.3)$$

where t is the daily exposure time in seconds.

For more than one exposure period:

$$\mathrm{eVDV} = 1.4 \times \sqrt[4]{a_1^4 \times t_1 + a_2^4 \times t_2 + a_3^4 \times t_3 \ldots}\ \mathrm{ms}^{-1.75} \qquad (14.4)$$

where a_1, t_1, etc., have the same meanings as before.

Calculation of VDV from a shorter VDV measurement

The vibration dose value for the shift can be calculated from the VDV measured over a shorter period, as long as this period is specified (when working with rms measurements it was not necessary to know the actual measurements period)

$$\mathrm{VDV}_{\mathrm{shift}} = \mathrm{VDV}_{\mathrm{measured}} \times \sqrt[4]{\frac{T}{t}}\ \mathrm{ms}^{-1.75} \qquad (14.5)$$

where $\mathrm{VDV}_{\mathrm{measured}}$ is the VDV from a shorter exposure; t is the time over which $\mathrm{VDV}_{\mathrm{measured}}$; $\mathrm{VDV}_{\mathrm{shift}}$ is the VDV for the whole shift; and T is the daily exposure duration.

For exposure to more than one period of vibration at different levels, the best approach is to work out the partial VDV for each individual period of exposure using equation 14.5. These partial VDVs can then be combined as follows:

$$\mathrm{VDV}_{\mathrm{combined}} = \sqrt[4]{\mathrm{VDV}_1^4 + \mathrm{VDV}_2^4 + \mathrm{VDV}_3^4 \ldots} \qquad (14.6)$$

where VDV_1, VDV_2, etc., are the respective VDV values assessed for each individual period of exposure.

Example

As an example, we will look at the exposure values calculated from exposure lasting 30 min to vibration which is measured over that period at 1.2 ms^{-2} using rms averaging equipment and for which a VDV of 11.7 $ms^{-1.75}$ is measured using rmq averaging equipment.

Equivalent 8-h exposure A(8):

$$A(8) = a \times \sqrt{\frac{t}{8}} = 1.2 \times \sqrt{\frac{30/60}{8}} = 0.3\,ms^{-2}$$

Estimated vibration dose value:

$$eVDV = 1.4 \times a \times \sqrt[4]{t} = 1.4 \times 1.2 \times \sqrt[4]{30 \times 60} = 10.9\,ms^{-1.75}$$

As the measurement was made over the entire daily exposure period of 30 min, the VDV measured is the same as the daily VDV, 11.7 $ms^{-1.75}$. The eVDV and VDV values here agree reasonably well. Because a short exposure time is involved, though, the VDV method is much more stringent. The A(8) is well below the exposure action value, but the VDV and eVDV are both above the corresponding EAV.

Case study 14.1 Assessing whole body vibration exposure

Pedestrian areas, car parks and access roads at the Fennyshaw Shopping Centre are kept clean by one operator using a hand-guided sweeping machine which has an add-on seat. The seat incorporates very little vibration isolation and the vibration exposure of the operator is of concern to the centre's health and safety manager. The machine is garaged at premises about half a mile from the centre and the daily routine is to drive it to the centre, and to spend an average of 6 h sweeping the centre's open spaces before returning to the garage and performing any necessary cleaning and maintenance work on the machine itself. Of the 6 h spent at the centre, it was determined by observation that 4 are spent driving the machine, and the remaining 2 on related activities not involving vibration exposure. The normal operating speed has been measured at 3 mph, while the top speed, used for travelling to and from the site, is nearly twice as fast as this.

Measurements were made of the whole body vibration on the machine's seat using a triaxial seat pad accelerometer. The measurements covered 1 h of normal cleaning operations and also – as a separate measurement – one journey from the garage to the shopping centre. The equipment used measured the rms acceleration along all three axes simultaneously – it was considered that this was adequate as there were no obvious shocks present in the vibration as felt by the operator. The results were as follows:

Axis	Sweeping	Travelling
x	$0.746\,ms^{-2}$	$3.24\,ms^{-2}$
y	$0.438\,ms^{-2}$	$3.13\,ms^{-2}$
z	$0.783\,ms^{-2}$	$1.51\,ms^{-2}$

The much higher acceleration levels measured while travelling at top speed were consistent with the operator's subjective impressions.

The equivalent 8-h acceleration A(8) was calculated for each axis, using assumed exposure times of 4 h for sweeping and 12 min for travelling.

x-axis:

$$A(8) = \sqrt{\frac{t_1 \times a_1^2 + t_2 \times a_2^2}{8}} = \sqrt{\frac{4 \times 0.746^2 + 0.2 \times 3.24^2}{8}} = 0.74\,ms^{-2}$$

y-axis:

$$A(8) = \sqrt{\frac{t_1 \times a_1^2 + t_2 \times a_2^2}{8}} = \sqrt{\frac{4 \times 0.438^2 + 0.2 \times 3.13^2}{8}} = 0.58\,ms^{-2}$$

z-axis:

$$A(8) = \sqrt{\frac{t_1 \times a_1^2 + t_2 \times a_2^2}{8}} = \sqrt{\frac{4 \times 0.783^2 + 0.2 \times 1.51^2}{8}} = 0.60\,ms^{-2}$$

When assessing the effects of whole body vibration on health, ISO 2631 requires the values for the x- and y-axis to be multiplied by a factor of 1.4. After this is done, the resulting A(8) values are as follows:

Axis	A(8)
x	$1.04\,ms^{-2}$
y	$0.81\,ms^{-2}$
z	$0.60\,ms^{-2}$

Using the exposure action and limit values for equivalent 8-h exposures, this means that the operator seems to be approaching the exposure limit value due to the x-axis exposure. Measured values for the other axes are both above the exposure action value.

Because of the short period of higher vibration exposure it was thought that the situation would be worse if the vibration dose value was calculated. A true

VDV result cannot be calculated from values measured with equipment which uses rms averaging. However, as there are no significant shocks present, an estimated vibration dose value or eVDV can be calculated for comparison with the VDV-based action and limit values.

x-axis:

$$\begin{aligned} \text{eVDV} &= 1.4 \times \sqrt[4]{t_1 \times a_1^4 + t_2 \times a_2^4} = 1.4 \times \sqrt[4]{14\,400 \times 0.746^4 + 720 \times 3.24^4} \\ &= 23.8\,\text{ms}^{-1.75} \end{aligned}$$

y-axis:

$$\begin{aligned} \text{eVDV} &= 1.4 \times \sqrt[4]{t_1 \times a_1^4 + t_2 \times a_2^4} = 1.4 \times \sqrt[4]{14\,400 \times 0.438^4 + 720 \times 3.13^4} \\ &= 22.7\,\text{ms}^{-1.75} \end{aligned}$$

z-axis:

$$\begin{aligned} \text{eVDV} &= 1.4 \times \sqrt[4]{t_1 \times a_1^4 + t_2 \times a_2^4} = 1.4 \times \sqrt[4]{14\,400 \times 0.783^4 + 720 \times 1.51^4} \\ &= 13.7\,\text{ms}^{-1.75} \end{aligned}$$

The x- and y-axis values once again need to be multiplied by 1.4 as required in ISO 2631 (it is a coincidence here that the same factor of 1.4 arises in the eVDV calculation and is also applied to the x- and y-axes). The final eVDV values are:

Axis	eVDV
x	33.3 $\text{ms}^{-1.75}$
y	31.8 $\text{ms}^{-1.75}$
z	13.7 $\text{ms}^{-1.75}$

These results confirm the initial assumption that the VDV would deal more harshly with a short period of significantly higher vibration exposure. This is further emphasized if the VDV resulting from (i) just the 6 h of sweeping; and (ii) just the 12 min of travelling are compared:

Axis	Overall eVDV	eVDV due to sweeping alone	eVDV due to travelling alone
x	33.3 $\text{ms}^{-1.75}$	11.4 $\text{ms}^{-1.75}$	32.9 $\text{ms}^{-1.75}$
y	31.8 $\text{ms}^{-1.75}$	9.4 $\text{ms}^{-1.75}$	31.8 $\text{ms}^{-1.75}$
z	13.7 $\text{ms}^{-1.75}$	12.0 $\text{ms}^{-1.75}$	10.9 $\text{ms}^{-1.75}$

Using A(8), the whole body vibration exposure is above the exposure action value (due to all three axes) and in one axis – the x-axis – is uncomfortably close to the exposure limit value in view of the measurement uncertainties which must be taken into account. Using the VDV method, the exposure is well above the exposure limit value on both the x- and y-axes, and is also above the exposure action value on the z-axis.

However the data are analysed, the conclusion seems to be that ways of reducing vibration exposure should be investigated urgently, and that the journey to and from the garage is the most damaging part of the daily vibration exposure. To reduce this – perhaps by imposing a speed limit – is the first priority.

Reducing Noise and Vibration Risks

Managing controls on noise and vibration exposure

The costs of noise and vibration exposure

Measures to control the noise and vibration exposure of employees are often expensive, and sometimes they can be very costly indeed. Failure to control these risks can also be very costly. The problem sometimes is that the costs of taking action to reduce risks are likely to be immediate, obvious and easily quantifiable. The possible costs of failing to take action are more remote, less certain, and less clearly attributable to one particular failure. Whether working in the private or the public sector, a financial manager has the overriding objective of keeping his organization's financial position within the budgetary targets set. There is normally no shortage of projects which could be financed, and it is not unusual for their proponents to argue that spending money on a particular project now will in the long term improve the financial position of the organization as a whole.

As an example, assume that an assessment of hand–arm vibration exposure has identified a problem in one particular department. The tools in use there are exposing their operators to high levels of vibration, and an exhaustive investigation has shown that if they were all replaced (at a cost of several tens of thousands of pounds) the hand–arm vibration exposure of these employees would be drastically reduced. The retooling would also lead to a small improvement in productivity. The existing tools, though, have several years' life left in them and a certain amount of retraining will be necessary if they are all replaced in the near future. The immediate costs and benefits are easily identified and short-term costs can be attached to both the possible courses of action.

In the longer term, things are much less clear. It is possible that if no action is taken a number of employees will develop HAVS symptoms. They are likely to take more time off sick, and some may eventually be unable to remain in the job. A few years from now, they may decide to sue the company for compensation. Compensation will be covered by the company's insurers, but the insurance company is likely to take various actions in response to a civil action. They may insist on changes in working practices, increase premiums, refuse to renew cover, or they may take some combination of these measures. HAVS is a notifiable disease, and the Health and Safety Executive may become involved if there is evidence that a number of cases are arising within the company. They could prosecute, or require immediate changes to working practice rather than the more gradual changes which result when action is taken under the control of the employer. None of these costs are inevitable, and none of them are easy to quantify. The employees concerned may be slow to develop HAVS symptoms or they may do so after they have left their present employment. Technical developments may make the process concerned obsolete or even lower vibration tools may become available within a short period. There may even be other safety concerns which require expenditure, and which are regarded as even more pressing.

It is not surprising that health and safety professionals are normally keen to invest in safer working arrangements, but neither is it surprising that they sometimes have to struggle to make their case.

When devising a programme to manage risks such as those of noise and vibration exposure, it is necessary first to identify the various control measures which are available, and to assess their likely effectiveness as well as their probable cost. At this stage various courses of action can be proposed. Normally, for example, it will be necessary to consider various combinations of engineering controls, revised management procedures and arrangements for medical surveillance. Then it will be necessary to identify the costs and the benefits which attach to each possible course of action. Some of these costs and benefits can be expressed explicitly in financial terms, but others will be less easy to quantify. For example, a company which is concerned for the health of its employees is likely to benefit from better industrial relations and lower sickness rates than one which is not. These factors in turn may result in improved ability to meet contract deadlines and hence in better success in gaining future contracts. Some public sector organizations, and increasingly private sector ones as well, audit the health and safety arrangements of companies bidding for contracts. Therefore, there may be commercial benefits in being able to demonstrate that risks such as these are being actively managed.

Insured and noninsured costs

Although it is a legal requirement in the UK to carry insurance for third party compensation claims, there are many costs involved in defending such a claim which are not normally covered by the policy. The time spent by managers and administrative staff in accessing records, making documents available and briefing lawyers is unlikely to be covered. The potential loss of business following a compensation claim which gains local publicity, and the need to train replacements for skilled and experienced workers who have to retire prematurely, or move to less skilled work, are examples of costs which are less easy to attribute to one particular cause.

The cost of arranging the insurance cover itself is also an uninsured cost. Insurance premiums are determined on the basis of the insurance company's perception of the risks to which it is exposing itself by providing cover. This perception may result mainly from a consideration of the type of business in which the organization is engaged, but insurance companies are becoming more sophisticated in assessing potential risks. It is normal for an insurer to arrange periodic health and safety inspections of a client's premises. If a generic health and safety inspection reveals a potential problem with noise and/or vibration exposure, then the insurer may insist on a specialist external assessment of noise and vibration risks – and on any remedial action that is proposed as a result – if cover is to be continued.

A hierarchy of control

When a noise and/or vibration risk is identified, the appropriate control measures will depend very much on the precise situation in which it has arisen. It is possible, though, to classify control measures in a more general way. This has the benefits of ensuring that all the possible approaches have been considered, and that the correct priority has been attached to the various control measures. One version of the hierarchy is shown in Table 15.1, and the various classes of control measure are discussed more fully below.

Eliminate the process

Many finishing processes are among those which inevitably expose operators to high levels of noise and/or vibration, and frequently alternative finishes can be considered. It is difficult to imagine that many employers would pay workers to

Table 15.1 A hierarchy of control

1	Eliminate the process
2	Substitute a less hazardous process
3	Outsource hazardous processes
4	Select machinery with reduced emissions
5	Maintain tools and equipment
6	Interrupt the path from source to receiver
7	Reduce exposure times
8	Supply personal protective equipment (PPE)

carry out operations which were completely unnecessary, even if no health hazards were involved. As technology advances and customer requirements change, though, it sometimes happens that traditional ways of working become obsolete. For example, a component which is to be painted does not require the high degree of finish which was normal before the paint process was introduced.

Many finishing processes are necessary because of defects in 'upstream' processes. For example, flash – waste material – often has to be ground off castings. If the quality of the casting process can be improved, then it may be possible to reduce the amount of grinding required, or even eliminate this task altogether.

Substitute a less hazardous process

Sometimes the same result can be achieved by using a completely different tool or process. For example, instead of using a hand-held concrete breaker, an attachment on a digging machine may do the same job while exposing the operator to much lower levels of hand–arm (but possibly higher levels of whole-body) vibration.

Outsource hazardous processes

A process which exposes employees to particular levels of noise or vibration could be contracted out to another company. This may absolve the employer of responsibility for the health effects (although in the case of contractors working on the employer's premises, this is not guaranteed). However, there may be ethical issues involved. The contractor who is prepared to take on this hazardous work could be a specialist firm who concentrate on this particular process. As a result, they have been able to invest in expensive plant, with reduced noise and vibration emissions, which a less specialized firm could not afford. Or it may just be that they are prepared to take on hazardous work because they do not attach a great deal of importance to the health and safety of their employees.

An increasing number of organizations, especially in the public sector, would not want to award a contract to a company falling into the latter category. This might result from a sense of social responsibility, but it could also come down to self-interest. A company which suffers from a high level of employee sickness is unlikely to have a good record of meeting delivery deadlines, particularly if their health and safety practices expose them to possible enforcement action from the authorities.

Select machinery with reduced emissions

Where it is decided that a particular process has to be carried out in-house, and by a particular method, physical hazards can be minimized by choosing machinery which is known to emit low levels of noise and vibration compared with other tools of the same type. More detailed guidance on choosing low emission tools appears later in this chapter.

Maintain tools and equipment

As a general rule, well-maintained tools have lower emissions of both noise and vibration. Maintenance in this context can include a number of considerations. Workers can be trained to keep machinery clean and lubricated and to replace cutting and abrasive components at an early stage before noise and vibration levels increase rather than when it becomes apparent that their performance has deteriorated. Some tools need regular attention. Abrasive wheels, for example, need to be redressed at frequent intervals using a special tool if their vibration emissions are to be minimized and this too depends on having a well-trained workforce. Frequently the quality and efficiency of the operation will also be improved by regular maintenance operations such as these. The timely carrying out of more fundamental maintenance will depend on regular inspections and servicing by qualified repairers, and of a system of reporting faults and problems as they arise in between more regular inspections.

Many tools' emissions of noise and, especially, vibration will increase dramatically if the tool is loaded beyond its design capacity. The life of the tool will also be reduced if this happens. Operators need to be aware of the capabilities of their tools and to be able to choose appropriate machinery for each task. They may also need training in appropriate techniques to minimize noise and vibration emissions. Managers equally need to be receptive to suggestions and comments from the workforce about the equipment provided.

Interrupt the path from source to receiver

A variety of measures is available to limit the transfer of noise from a source to the human beings who may be affected by it. These are discussed more fully in the next chapter. In the case of vibration, the operator is normally in direct contact with the source of vibration energy and opportunities for interrupting this path are few.

Reduce exposure times

The daily dose of both noise and vibration will depend both on the noise and vibration emissions of the equipment in use, and on the time for which each worker is exposed to them during any particular day. In the case of hand–arm vibration it is a common outcome of measurements that a daily time limit is set for the use of that tool. Time limits may also arise from a noise assessment. It may be possible to limit the total amount of work carried out on a particular process, failing which arrangements can be made to share the necessary work between two or more employees. It is important that any time limits set take account of exposure to noise and/or vibration from other tools during the same shift.

Personal protective equipment (PPE)

It is a truism that PPE should be the last resort when controlling any workplace hazard. In the sense that alternative methods are almost always preferable, this cannot be denied. In a strict chronological sense, though, PPE often comes first. It can be brought into use within a few hours of a problem being identified, whereas the other approaches listed here will normally take considerably longer to plan and implement.

Hearing protection is discussed in detail in Chapter 17. Correctly specified and consistently used, it can have the effect of reducing noise doses by 30 decibels. In terms of the sound energy received by a human ear, this is equivalent to a 1000-fold reduction.

Against this must be set the difficulties of ensuring that it really is properly specified and implemented, and the consideration that hearing protection can only benefit the wearer while many other measures can be relied on to reduce the noise exposure of a larger number of people. Many workers are reluctant to use hearing protection because it can be uncomfortable and because it interferes with communication. A further issue is that of possible negative interactions between the hearing protection and measures to control other workplace hazards. An example of this is the fact that hearing protection wearers may not be able to hear the fire alarm.

Although personal protective equipment can sometimes be very effective against noise, the situation with vibration is very different. The antivibration gloves on the market do not significantly reduce hand–arm vibration exposure. No personal protection at all is available to reduce whole-body vibration exposure.

The specification of low emission machinery

Substitution of a less hazardous process is one of the actions listed in the ‘hierarchy of control’ above. If there is no alternative process available, it may be necessary instead to reduce noise and/or vibration exposure by replacing plant and tools with alternatives which can be relied on to expose workers to lower levels of these hazards.

There may be a case for disposing immediately of a particularly hazardous piece of equipment, but machinery which is frequently used will have a limited life and will be replaced as a matter of course after a few years. The choice of new equipment and machinery will be decided by a number of factors including cost, efficiency and other safety considerations. However, noise and vibration emissions will be an important consideration whenever exposure of employees to these hazards is significant.

Under the 1992 Supply of Machinery (Safety) Regulations (which implement in the UK the EU Machinery Directive), suppliers of machinery have various duties in relation to the emissions by their products of both noise and vibration. One of these duties is to reduce emissions as far as possible, taking into account technical progress and the availability of means to reduce these emissions. Another is to provide information about any remaining emissions. This information will be provided on request and will also appear in the operating instructions of the equipment. Some sort of measurement of noise and/or vibration emissions will be necessary. The precise information which is required, and the necessity for further measurements, depend on the outcome. Table 15.2 lists the information required.

In these regulations, machinery is defined very widely, but there are some specific exceptions, including:

- Vehicles (other than those used in the mineral extraction industry)
- Machinery for medical use or for military or police use
- Fairground and amusement park equipment
- Steam boilers, tanks and pressure vessels
- Firearms
- Storage tanks and pipelines for petrol, diesel fuel, inflammable liquids and dangerous substances.

Although the regulations apply to machinery for use in workplaces, it is common for machinery intended for other uses to find its way into a workplace. Many

Table 15.2 Information to be supplied under the Machinery Directive

Hazard	Condition	Information required
Noise	If the A weighted continuous sound pressure level at the operator's position is less than 70 dB	A statement that the A weighted continuous sound pressure level at the operator's position is less than 70 dB
	If the A weighted continuous sound pressure level at the operator's position is between 70 and 85 dB	The measured value of the A weighted continuous sound pressure level at the operator's position
	If the A weighted continuous sound pressure level at the operator's position is greater than 85 dB	The A weighted sound power level of the machine
	If the C weighted peak sound pressure level at the operator's position is greater than 63 Pa (130 dB)	The measured value of the C weighted peak sound pressure level at the operator's position
Hand–arm vibration (applies to hand-held and hand-guided machinery only)	If the hand–arm weighted rms acceleration is less than 2.5 ms^{-2}	A statement that the hand–arm weighted rms acceleration is less than 2.5 ms^{-2}
	If the hand–arm weighted rms acceleration is greater than 2.5 ms^{-2}	The measured value of the hand–arm weighted rms acceleration
Whole body vibration (applies to mobile machinery only)	If the weighted rms acceleration is less than 0.5 ms^{-2}	A statement that the weighted rms acceleration is less than 0.5 ms^{-2}
	If the weighted rms acceleration is greater than 0.5 ms^{-2}	The measured value of the weighted rms acceleration

suppliers, therefore, err on the side of making information available. The duty to provide the information lies with the manufacturer if it is manufactured within the European Union, or with the company that first imports it into the EU.

For noise emissions, the initial sound measurement is of the sound pressure level at the operator's position. In cases where there is not an obvious operator's position, a standard position is defined. If sound power measurements are required, these will normally need to be carried out in a special acoustic test chamber, or to be made with the assistance of advanced equipment such as a sound intensity analyser.

The 2001 Noise Emissions in the Environment by Equipment for Use Outdoors Regulations also require suppliers of various types of machinery to measure and declare sound power levels of their products. This piece of legislation implements a different EU directive which is intended to bring about a reduction in environmental noise levels. As well as requiring tests and the declaration of sound power levels on a wide range of outdoor equipment, these regulations specify that the machinery covered must be clearly labelled with its measured sound power level. For a more restricted class of outdoor machines, there is a limit on the permissible sound power level. In some cases these sound power level limits vary according to the power rating of the individual model concerned. Although not intended mainly to control workplace noise exposure, the information supplied under these regulations can be a useful addition when managing the noise exposure of outdoor workers.

There are various international standards available (for example, ISO 3740-6 and ISO 9613) which specify the various methods for measuring sound power levels in the same way that ISO 5349 describes how hand–arm vibration should be measured at a tool handle for the purposes of assessing human exposure. The actual emissions of both noise and vibration, though, will depend on the way a machine is operated, the materials it is worked with and in some cases on factors such as the operating speed. For example, a circular saw running freely will be less noisy than one being used to cut MDF panels, while a badly maintained grinder will expose the operator to more vibration than a brand new one. Useful comparisons between machines can only be made if they are operated in a similar way at the time of the measurements, and to assist with this a number of test codes have been produced. When these test codes are written, the main priority is to establish a reproducible procedure. So far as possible it should also be one which realistically represents its use in a typical workplace (though not, of course, any foreseeable workplace). Most test codes are reasonably reproducible, and some, but not all, are also realistic in terms of operating conditions in some workplaces. The ISO 8662 series of standards, for example, deals exclusively with the measurement of hand–arm vibration at the handles of power tools, and especially pneumatically powered tools. Each part relates these measurements to a different class of tool. BS 6916 is an example of a standard which takes a different approach. This series of standards covers different aspects of the safety of hand-held chain saws and Part 8 of this standard deals with hand–arm vibration. Table 15.3 illustrates the type of detail specified in these test codes

Table 15.3 Some features of two test codes

	ISO 8662 Part 4:1995	BS 6916 Part 8:1988
Type of tool covered	Hand-held power grinders	Portable chain saws
Measurement direction	z-axis weighted acceleration	Weighted acceleration along all three axes
Quantity to be specified	z-axis weighted acceleration	Root-sum-of-squares weighted acceleration
Handles to be measured	Both	Both
Accelerometer position	Shown diagrammatically for four types of tool	Shown diagrammatically for a single design of handle
Mounting of accelerometers	Three alternative methods described	One method described
Tool condition	New, serviced and lubricated	New, serviced and lubricated. Engine warmed up
Object to be cut	None. The grinding wheel is replaced with an unbalanced aluminium wheel	A test log is described in the standard. Dimensions depend on the size and power of the saw
Operating speed	Nominal load speed	Three speeds defined
Feed force	A specified force is applied to a specified point on the tool, but no material is worked during this test	Not specified

– and the variations that can occur between them – by comparing some details of ISO 8662 Part 4 and BS 6916 Part 8.

Because of the sometimes artificial conditions under which these measurements are made, it is not normally possible to take a figure from the supplier's data – whether of noise or vibration – and assume that that is the level to which the operator will be exposed in a real workplace. These figures are not intended to be used in this way, but aim instead to make easy comparisons between different tools and machines which could be used to do the same job.

Hewitt and Smeatham (2000) studied the test codes for hand–arm vibration emissions of seven classes of tool. They conclude that 'The standards investigated are generally repeatable and reproducible and ... produce data that indicates the relative risks between tools, within broad margins. However, in general, the vibration magnitude measured using the standard test did not give a reliable indication of the risk of using the tool in practice.' Two of the seven test codes considered, though, were not found even to rank tools reliably in the correct order.

Because of these considerations, a single figure quoted as either a noise or vibration emission value is not adequate. At the very least, the test code used must be specified along with the quantity measured and the values obtained. If the test code allows alternatives at any point, a full specification should show which alternative was chosen, and if there is no published test code for that type of machinery a detailed description of the test conditions used should be included. Increasingly, test codes are emphasizing the importance of estimating and declaring the inherent uncertainty in any measurements – called the 'K value' – and in that case this information should be included too. 'Noise,

92 dB' is not a sufficient statement of noise emissions. It is not even clear whether this value is the sound pressure level or the sound power level.

Using health surveillance information

Most large companies and organizations have an in-house occupational health service, while smaller companies normally use an external service. There can sometimes be a lack of effective communication between occupational health professionals on the one hand and health and safety managers on the other. At its worst, such a lack of communication can lead to the occupational health department collecting information which charts the progress of damage to employee's health, while those responsible for the management of health and safety fail to intervene to prevent it. This is good neither for employees' health and welfare nor for the long-term financial prospects of the employer

An effective occupational health service will raise the alarm if there seems to be any deterioration in employees' health, including the effects of noise and vibration exposure. To make this possible it is necessary to carry out baseline surveillance on new employees, and to target this initial surveillance effectively it will be necessary to have an assessment of any significant hazards to which the employee could potentially be exposed, such as the use of vibrating tools or work in noisy environments. Although mainly intended to establish a reference point to identify future deterioration in health, baseline surveillance has an additional attraction for employers (and their insurers) of identifying any existing damage for which the present employer then cannot be held liable in the future.

If health surveillance seems to indicate that an employee or a group of employees are developing, for example, the symptoms of HAVS, or significant levels of hearing loss, then it will normally be necessary to reassess their exposure and the effectiveness of any control measures in place. It may be that the health effects can be limited by appropriate training, by more effective hearing protection, or by improving management procedures.

Reviewing noise and vibration exposure control programmes

Large organizations will normally operate under procedures which automatically initiate reviews of their hazard control programmes. An assessment of noise and/or vibration exposure will normally result in the adoption of a number of control measures, along with the specification of a period after which their implementation and effectiveness will be reviewed. One of the measures included should be a reassessment of exposure either when there are significant changes in working practices or after an appropriate period to allow for gradual changes. A review of the programme as a whole will go further than this. It will, for example, include an examination of the reassessment interval (it may be appropriate to start by reassessing as soon as the control programme is implemented, but thereafter to reduce the frequency). It will also involve a critical evaluation of the effectiveness of the control measures in the light of the experience of operating them.

Case study 15.1 The use of emission data

Hangingstone Landscapes specializes in landscaping and grounds maintenance. The work is very seasonal and power tools, such as brush saws and vacuum/ blowers, can be in use for long periods at certain times of the year. Following a survey by an external consultant of hand–arm vibration exposure it was decided that some tools, including strimmers and chain saws should be replaced with models which would result in reduced hand–arm vibration exposure.

A short-list was drawn up from suppliers' literature of models which were capable of carrying out all the required work. A maximum cost for each type of tool was also decided at this point. The relevant manufacturers were asked for vibration emission data. In some cases, this was already available from catalogues and in others it was readily forthcoming when requested. In one case, data were eventually supplied after several requests.

The data supplied were used to select the four models in each case which seemed to have the lowest vibration emissions. However, it was noted that all the data provided in this way were derived from laboratory tests which might not be very representative of the vibration emissions in practice. Some time was spent searching on-line databases for in-use vibration values, but this search was not very successful. The required data were positively identified only for one model of chain saw. For a further chain saw, and for one model of strimmer, possibly relevant data were discovered, but this was not certain due to confusing model numbers.

Manufacturers and distributors were contacted and were willing in most cases either to lend a machine for evaluation or to demonstrate one on site. With the help of the consultant who had carried out the original hand–arm vibration exposure assessment, hand–arm weighted accelerations were measured at the handles of each of the short-listed tools. It turned out that the lowest vibration measurements were indeed on the handles of the chain saw for which the laboratory test data were also lowest (although the actual acceleration measured was more than 70 per cent higher than the quoted figure). In the case of the strimmers, the tool with the lowest measured acceleration in the field trials was the one with the second highest declared value.

After further checks on the suitability of the selected tools, an order was placed for the tools which had shown the lowest vibration exposure in practice. The vibration exposure assessment was brought up to date using the measured acceleration values on the new tools, and as a result time limits on their use could be relaxed while still achieving a reduction in the assessed 8-h equivalent exposure of the operators. It was decided that HAV exposure would be reassessed and tool purchasing policy would be reviewed in 3 years' time.

Noise control

Approaches to noise control

When an employee is exposed above the second or peak action levels of the 1989 regulations the employer has a duty to reduce noise exposure to the lowest level reasonably practicable by means other than the use of hearing protection. At any level of exposure, the employer has a duty to reduce, so far as is reasonably practicable, the risk of hearing damage. There may be other reasons why it is desirable to reduce noise levels in a workplace. For example, it may improve efficiency if employees can communicate more easily with each other and use telephones and other devices. Or there may be other safety issues such as the audibility of alarms.

The detailed design of noise control systems involves principles and techniques well beyond the scope of this book. Normally they would not fall within the responsibility of those managing exposure to noise hazards. This chapter aims to cover the main principles involved in sufficient detail to help readers to:

- Suggest to employers a range of possible approaches to noise reduction which should be investigated further with those who have detailed knowledge of the operational, engineering and financial issues involved.
- Evaluate proposals for noise control submitted by specialist consultants and contractors.

Noise in workplaces is frequently radiated by vibrating surfaces. The control of this vibration can be an important step in controlling noise levels. When considering vibration exposure and the ways in which it can be controlled, many of the principles covered in this chapter are also very relevant to the reduction of human exposure to both whole body and hand–arm vibration, and a basic understanding of the principles of vibration isolation can be useful in evaluating the effectiveness of various approaches to the reduction of vibration magnitudes. However, reducing exposure to vibration from hand tools or from vehicles involves a number of complex engineering issues. It will only rarely be the case that vibration exposure can be reduced by modifications carried out within a workplace. Reducing vibration as a method of noise control, though, can much more often be achieved by those armed with a basic understanding of vibration control principles and a willingness to experiment and evaluate alternatives. Some of the principles of vibration control are therefore included in this chapter, but will be referred to in Chapters 18 and 19.

As a way of prioritizing the various alternative approaches to noise control, it is normal to consider them under three headings:

1. Noise control at source
2. Noise control on the path from source to reception point
3. Noise control at the reception point.

This is also the preferred order in which to apply the various available noise control measures. Most noise sources will affect a number of exposed individuals, and the noise energy is likely to travel to each affected individual via more than one route. Reducing the noise emitted by the source has the distinct advantages of reducing the exposure of every individual, and of reducing the noise arriving via each of the possible paths. However, it is not always clear whether a particular noise control measure should be regarded as a 'source' or a 'path' measure. It will, however, be preferable to tackle noise earlier rather than later in its path to the reception point, even if reduction at source is not practicable.

Reduction of noise at the reception point (i.e. the exposed individual) is normally tackled by means of hearing protection, which is the subject of Chapter 17. An alternative in some cases is the construction of a noise refuge, and some of the issues involved in the design of these are discussed later in this chapter.

In most cases, more than one noise source is present. First it is important to establish the relative importance of the various sources. This can sometimes be done by switching sources on and off selectively to establish which make the greatest contribution. If this is not possible – sometimes machines necessarily operate together or are connected by interlocks for safety reasons – then the relative contributions can be estimated by calculation. Noise reduction efforts need to be directed first to the source making the greatest contribution, as is illustrated in Case study 16.1.

Some approaches to controlling noise at source

Measures which can be used effectively to reduce noise levels at source will depend on the particular source concerned. Source reduction may be controversial in some cases where the noise is generated deliberately. It is, for example, normally possible to reduce the noise exposure of workers in entertainment establishments simply by turning down the volume control, but as the noise concerned is the purpose of the establishment, it may not be considered desirable to do this.

Noise generated by impacts can be reduced in a number of ways. An example of this type of noise is where metal components fall off the end of a conveyor belt into a metal container. The noise arises in this case – and in similar situations – both from the initial impact and also from the vibration of the container after the impact. A reduction in the impact speed will reduce both types of noise, and this can sometimes be achieved simply by reducing the height through which the components fall – perhaps by standing the container on a plinth. If this is not practical, then a cushioning layer will have a similar effect on the noise, although this assumes that a suitable material can be found to line the container which will survive the impacts for a sufficient period to make this solution cost-effective. An appropriate lining on the outside of the container will be less liable to damage. It will not have any effect on the initial impact noise, but will act to damp the post-impact vibrations and therefore to reduce the overall noise emissions.

A similar problem can arise in sheet metal works, where many processes involve impacts between tools and sheets being worked. Once again, reduction of impact speeds and damping of sheet materials are possibilities. Metal work-benches can be replaced with other materials, or damping treatment can be applied to them. Sheets being cut or worked can have a vibration damping treatment applied before work starts.

Air, steam and gas discharges can generate a great deal of noise. Sometimes this results from the operation of a pressure release valve, but many pneumatic tools and clamps release air as part of their normal operation. Diffusers can be fitted to many discharge valves at relatively low cost which reduce noise levels significantly. If the discharge is of steam or of another gas, then the diffuser chosen must be one which is intended for that purpose. Noise emissions from these sources depend on the supply pressure and it may be possible to reduce this pressure to that which is strictly required rather than supplying all equipment at the higher pressure required by one machine. Helical saw and cutter blades are now standard on many machines. Each tooth on this type of blade is curved so that it engages with the work progressively rather than striking it suddenly. The helical blades result in lower noise emissions as well as reduced wear.

A siren is the name given to the type of noise source which works by periodic-ally interrupting a moving air stream. These are very efficient noise sources, and as a result have traditionally been used for emergency warning systems intended to be audible over a wide area. In the workplace they are less desirable. A siren can sometimes be created when a dust extraction system is added to an old machine, such as a wood planer, which was originally manufactured without any extraction. Without very careful design, it is possible for the air to be drawn through the table where it is chopped by the cutting blades. A weighted noise level in the region of 105 dB can be generated at the operator's position, mainly in one narrow band of frequencies. The solution is to redesign the airflow so that it is drawn in by another route.

Free field and reverberant field

When considering noise control measures, it is essential to be aware of the difference between direct and reverberant sound. At many indoor workstations, part of the noise to which the employee is exposed will come from the equipment being operated by that individual, and which is generated within a metre or two of his/her ear. This is the direct noise, and it travels straight from the source to the operator. It will be supplemented by noise from other sources within the room. Normally there will be a number of such sources at various positions, all of which are further away than the source of the direct sound. Together, these more distant sources make up the reverberant sound. Most of the sound energy from these sources will arrive at the position of interest via reflections from the various hard surfaces in the room, so that the reverberant sound arrives from all directions. It has another important characteristic. If the reverberant sound level is measured at various points around a small to medium workshop (this can be done by measuring at positions that are at least 2–3 m away from any identifiable noise source), the results are unlikely to vary by more than a few decibels.

In an outdoor situation there is normally no reverberant sound and all the noise arriving at the workstation travels directly from the source. It can also be expected to fall off sharply as the distance from the source increases. For a physically small noise source it can be assumed that doubling the source–receiver distance will result in a 6 dB fall in the noise level. In some outdoor situations, this can be the key to reducing noise exposure. If there is no need for an individual to work close to a noise source, then simply increasing the distance can reduce noise exposure significantly. Indoors, it would be necessary first to establish whether the main noise source is the direct sound from nearby machinery, or the reverberant sound field from more distant sources. In the latter case, increasing the distance will have very little effect. Other noise control measures which operate on the path from source to receiver will also be more or less effective depending on whether the main contribution to noise exposure comes from direct or reverberant sound.

Sound insulation

Sound insulation is the property of materials and structures which enables them to resist the transfer of sound energy from one side to the other. Contrary to what is sometimes believed, porous and fibrous materials are not good sound insulators – their role in noise control is a different one which is described below. Materials which are good sound insulators will either have a high surface density, or they will be more complex lightweight structures incorporating structural isolation. Thus a masonry wall made from good quality concrete blocks will be a good sound insulator. It is the structure most often used for party walls separating houses and flats. Multiplex cinemas are more likely to use plasterboard walls to separate their different auditoria, but to achieve good sound insulation these must be constructed in such a way that the two skins of the wall are completely separate. The more traditional partition wall consisting of two sheets of plasterboard supported on a single timber framework is a much poorer structure for sound insulation as it incorporates neither significant mass nor effective structure breaks. This kind of lightweight wall is adequate within buildings, but would not be suitable for separating two adjacent houses or flats.

Even if a good choice of materials and/or structures is made, good sound insulation will only result if gaps that allow sound energy to pass through the structure are avoided. The performance of a masonry wall will not even approach the expected value if large gaps are left in the mortar between the blocks (this sometimes happens in buildings as poor workmanship can be hidden by a coat of plaster). Inside a building, the sound insulation between two rooms can be decided almost entirely by any doors or other openings in the wall between them. The material of the door itself will almost certainly be a poorer sound insulator than the wall structure, but even more important is the fact that there are unavoidably gaps where the door joins its frame and where the frame joins the wall.

The principles of sound insulation are important when considering the design of noise reducing enclosures.

Sound absorption

When a sound wave strikes a solid surface such as a wall, the energy it contains is split into three parts:

- Some of the energy – normally a very small proportion – will travel through the wall.
- Some of the energy is reflected back into the room. There it helps to create the reverberant sound field which has been referred to earlier.
- Some of the energy will be turned into heat at the surface of the wall.

Different surfaces will reflect different proportions of the energy which strikes them. Furthermore, the proportion reflected will depend upon the frequency of the incident sound wave.

Generally speaking, rigid, nonporous surfaces, such as concrete or plastered blockwork, reflect a high proportion of the incident sound energy. These types of surface are common in industrial buildings.

Porous materials, such as foam and mineral wool, absorb high frequencies very effectively, but have little effect on low frequency sound. The frequency above which they begin to absorb sound effectively is related to the thickness of the material.

Panel materials are more effective at absorbing low frequency sound. This type of absorber consists of a thin sheet of material – such as plasterboard or plywood – mounted over an air gap. It is possible to design panel absorbers to absorb sound in the required manner, but very often these absorbers arise for reasons other than noise control. Partition walls, large windows and laminated roofs are all structures which can help to reduce low frequency noise levels.

Figure 16.1 shows how the absorption coefficient of and a porous absorber might vary with frequency and with thickness. Many industrial buildings are constructed largely of hard, nonporous materials and structures which will obviously encourage high reverberant noise levels. In some cases reducing this reverberant noise can be of considerable benefit in reducing employees' noise exposure, although it must be remembered that this benefit will not be felt by those working close to noise sources and affected mainly by the direct sound field. There are several reasons why sound reflecting materials are used so frequently. Cheapness and ruggedness are commonly important. The need to clean surfaces easily is also often a factor, and in some workplaces cleanliness and hygiene are paramount.

Porous absorbing materials have several disadvantages as compared with the surfaces described in the previous paragraph. They are frequently easily damaged by contact with passing human beings, materials and machines. They may shed harmful fibres, and in many environments the pores which give them their sound absorbing properties will become clogged with oil or dust. In food handling establishments they are unacceptable for hygiene reasons, and in humid or hot environments they may have a very limited life. There are ways of protecting the porous surface while maintaining its sound absorbing character. A perforated metal facing can prevent abrasion and a plastic membrane can be incorporated to make a porous absorber suitable for hygienic environments (Figure 16.2). These

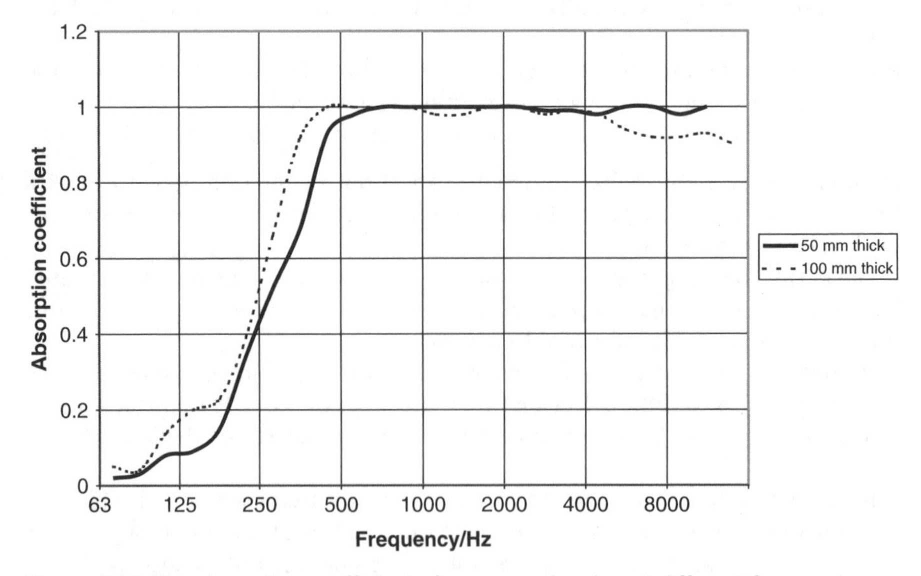

Figure 16.1 The absorption coefficient of a porous absorber at different frequencies.

Figure 16.2 A sound absorbing screen used in a factory.

more complicated absorbing structures tend to be expensive. Finally, to achieve effective absorption even in the middle of the frequency range, a considerable thickness of porous material is required (see Figure 16.1).

As a wall treatment, a sound absorber thick enough to be effective at low frequencies will occupy a great deal of valuable space. Porous sound absorbers are sometimes hung from the ceiling in reverberant industrial buildings. This protects them from physical damage – though probably not from dust – and uses space which would otherwise be wasted. It can be a useful way of controlling reverberant sound levels if this is the problem.

Vibration isolation

Radiation of sound from a source directly into the atmosphere is the most obvious, and normally the most important, route from source to receiver. Other pathways normally exist, and it may only be after measures have been taken to control the direct sound that the importance of these routes becomes apparent. Sound energy can be transmitted as waves travelling through solid materials, and is normally referred to as vibration when this happens – the precise relationship between sound and vibration lies outside the scope of this book.

Vibrational energy is transmitted effectively through rigid connections between different bodies. A thin vibrating panel of material can be an efficient

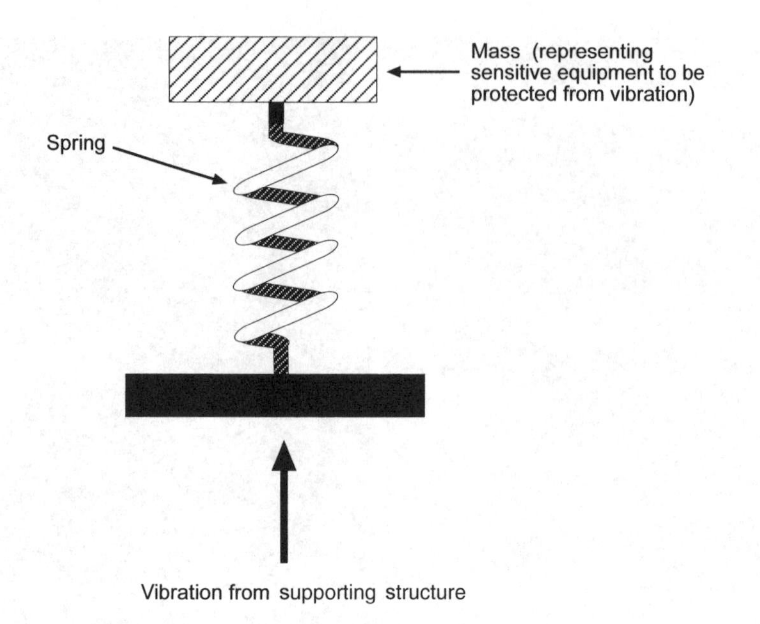

Figure 16.3 A simple vibration isolation system.

noise generator. It is no accident that a loudspeaker consists of just such a light vibrating panel. Vibrational energy transferred, perhaps by an intermediate structure, from a vibrating machine to a radiating panel can generate unexpectedly high noise levels. Occasionally, the panel may have a natural frequency which coincides with the frequency of the vibration. In this case, the phenomenon of resonance occurs and noise emissions are higher still. The solutions to vibration problems of this kind lie in isolating the source of vibration from possible radiating surfaces and/or in damping the vibration of this surface. A combination of measures may be required, but isolation close to the point of origin is the preferred approach as it affects all possible transmission paths.

To point out some of the benefits – and some of the potential pitfalls – when using vibration isolation systems, it is necessary to look at the behaviour of a very simple vibration isolation system as shown in Figure 16.3.

Imagine that a vibrational stimulus is applied to the base of the spring, while the resulting vibration of the suspended mass is monitored. The ratio of the amplitude of vibration of the mass to the amplitude of the support's vibration is known as the transmissibility of the system. Figure 16.4 shows how the transmissibility of the system varies with frequency. To interpret the graph it is necessary to bear in mind the following:

- The natural frequency of the system (labelled as f_0) is the frequency with which it would oscillate if disturbed from equilibrium (for example, if it were pushed slightly downwards and then released).

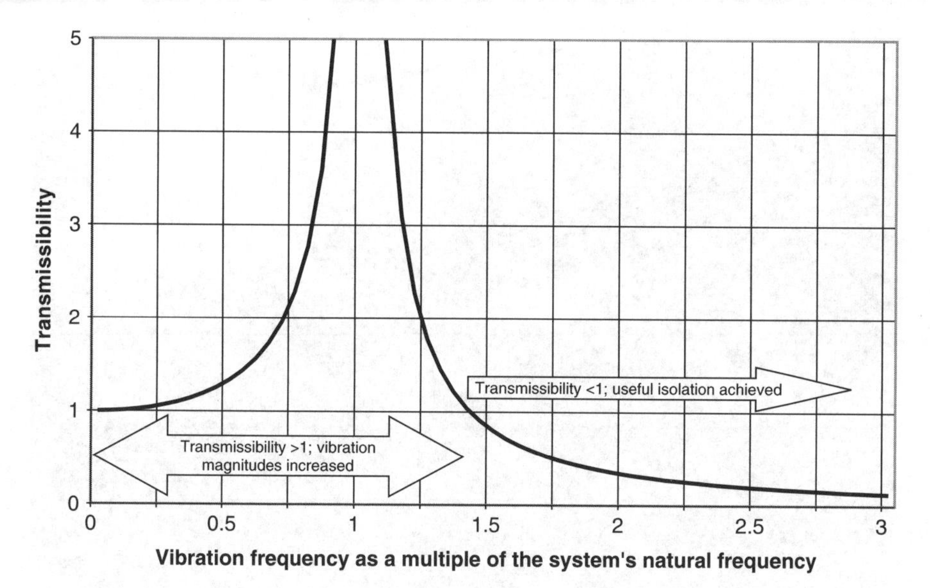

Figure 16.4 Variation of transmissibility with frequency.

- A transmissibility of one means that the isolation system is useless; the vibration of the mass is exactly the same as the vibration of the support.
- A transmissibility greater than one means that the vibration of the mass is greater than the vibration of the support.
- A transmissibility less than one means that the vibration of the suspended mass is less than that of the support.

Figure 16.4 shows that this simple isolation system has the following features:

- If the vibration applied has a frequency less than about 1.4 times the system's natural frequency, then the vibration of the mass to be isolated is actually greater than the vibration applied to the system.
- If the applied vibration is about the same as the natural frequency of the system, then the resulting vibration magnitude will be very large indeed. If there is no damping at all in the system it would be infinite, but in practice there is always some damping present.
- At frequencies greater than 1.4 times the natural frequency, the reduction in vibration magnitude achieved will increase steadily as the frequency of the applied vibration increases.

When additional damping is added to the system, either deliberately or in the course of making other modifications, then the extra damping has the following effects:

- Useful isolation is still achieved once the natural frequency of the system is less than 0.7 times the vibration to be isolated. However, the actual reduction in vibration magnitude falls as the degree of damping increases.

Figure 16.5 Vibration isolating machine mounts supporting a punch press.

- The magnitude of the peak in vibration magnitude, which occurs when the applied vibration is the same as the natural frequency of the system, falls as the degree of damping is increased.

Increasing the damping of the isolation system is therefore useful for preventing problems as a machine runs up to full speed. However, excess damping reduces the effectiveness of the isolation once it is running at its operating speed.

With this knowledge, it can be seen that there is great scope for making vibration problems worse as a result of an inexpert attempt to solve them. Unless the machine support system has a natural frequency lower than 0.7 times the problem frequency, then no benefit will result and in all probability the problem will become worse. If the natural frequency of the chosen system is close to the frequency of vibration, then the problems can become very much worse than before.

In this very brief discussion of the problems of vibration control, it remains to draw a distinction between the damping and the stiffening of a vibrating surface.

Imagine an engine cover plate is vibrating. No practical way of isolating the plate from the source of vibration energy can be found, but there are ways to limit the movement of the plate itself:

- Damping involves the conversion of vibrational motion into heat energy by applying a material of a type known as visco-elastic. This will normally reduce vibration magnitudes and hence radiated noise.

- Stiffening the plate – perhaps by welding a length of steel bar across it – appears to be an equivalent procedure. However, in this case no energy is absorbed. The main effect is to increase the natural frequency of the cover plate, and with a bit of luck this may reduce the amplitude of its vibration by moving it out of the resonant region. It is also possible that without an understanding of the process, stiffening the plate could increase its vibration amplitude by bringing its natural frequency closer to that of the vibration source.

Noise control on the transmission path

Once all that is practicable has been done to reduce sound emissions at source, it will be necessary to take steps to impede the transmission of sound to those positions where it will affect human beings. The precise strategy to be used will depend on the particular features of the individual workplace. Indoors, it could involve some combination of sound insulation, sound absorption and vibration isolation. Outdoors, there are two other strategies which may be of use:

- Increasing the distance between source and receiver can drastically reduce the levels to which employees are exposed.
- Barriers can be effective at reducing noise levels. A barrier is an obstacle such as a wall which does not completely enclose the source. Barriers are particularly effective at reducing high frequency sound, and to achieve a significant reduction it is important that there is no direct line of sight from the source to the reception point. The higher the barrier, the more effective it is likely to be. For a given height of barrier, it is normally more effective to position it close to the source rather than midway between source and receiver. Indoors, barriers are not normally very effective as sound is reflected off surfaces on both sides of the barrier so that it does not impede most of the possible paths for the transmission of sound. Figure 16.6 illustrates the use of barriers. However, coupled with appropriately placed absorbing materials, barriers can achieve a limited degree of noise reduction indoors.

Indoors, a variety of options will probably be available to control noise by interfering with the transmission path. It may be practical to enclose the most important noise sources once these have been identified, and some practical considerations relevant to the enclosure of machinery are discussed towards the end of this chapter. Sound absorbing surface treatments are most effective when applied close to the main noise sources rather than distributed around a large enclosed space. In this context, barriers and partial enclosures – apparently dismissed above – are not completely useless in reverberant spaces. The key to their use is to combine them with sound absorbing treatments on the barrier itself, and on walls and other surfaces close to the noise source. Useful noise

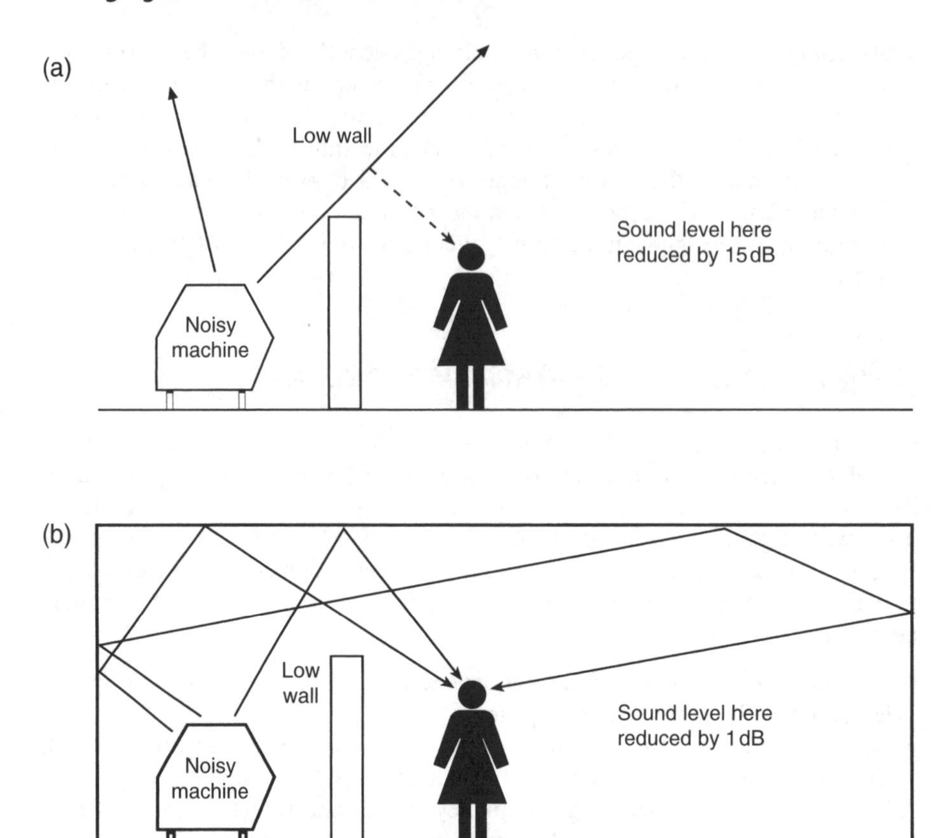

Figure 16.6 Barriers indoors and outdoors.

reduction can be managed in this way while maintaining easy access to the noisy machine itself. Figure 16.7 illustrates this principle.

Practical noise enclosures

Figure 16.8 illustrates the key features of a noise reducing enclosure. The key objective is to reduce the levels of noise to which persons in the vicinity are exposed to an acceptable level, while interfering as little as possible with the normal operation of the enclosed machine and those surrounding it. The three techniques involved are:

- Enclosing the source as completely as possible
- Absorbing sound within the enclosure, and
- Isolation of vibration from the exterior of the enclosure.

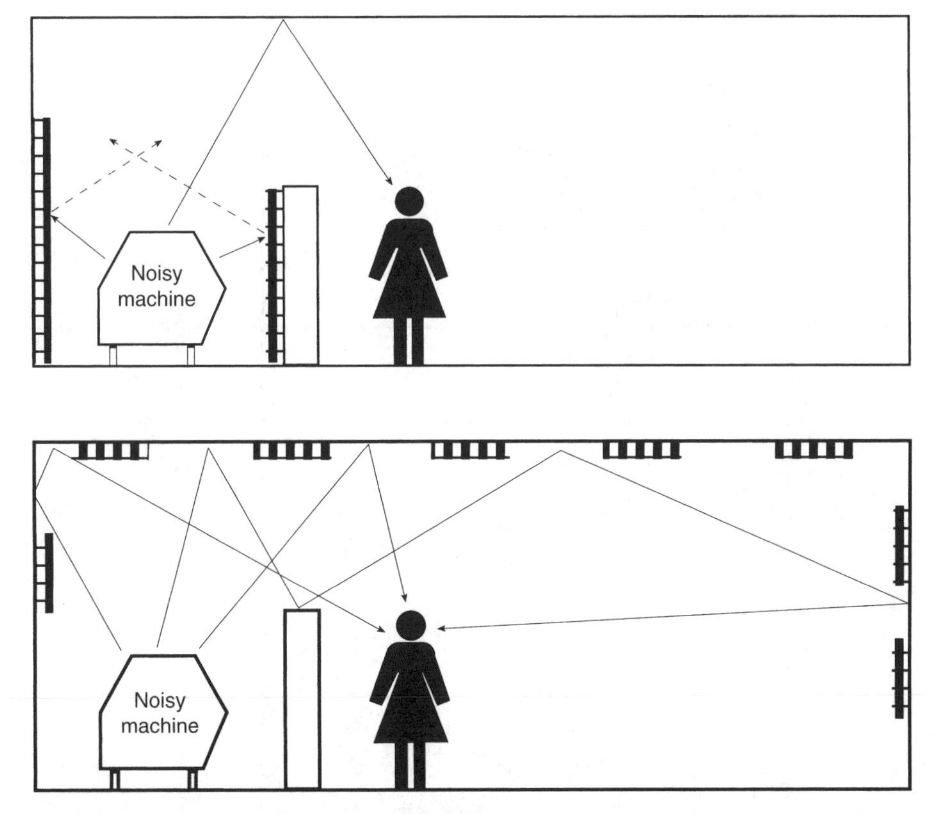

Figure 16.7 Effective use of sound absorption. The same area of sound absorption is more effective when concentrated around the main noise sources.

It is normally far more cost-effective to use all three of these techniques than to try to achieve the required noise reduction simply by using a more and more massive enclosure.

It is rarely possible to enclose a machine completely. In most cases it is necessary to feed materials into the machine and to remove finished work. This needs to be allowed for at the design stage so that this can be done while maintaining the effectiveness of the enclosure as a whole. Hurried modifications later on may have a serious effect on this effectiveness. Pipes and ducts for other services may also need to enter the enclosure, and it is necessary to allow for these. An enclosed machine may overheat even if without the enclosure no special cooling arrangements were necessary. Air supply and extract systems can be ducted outside the building (subject to environmental considerations) or if they are vented indoors they will need to be fitted with silencers.

Access for routine adjustments and occasional maintenance needs to be allowed for. If it is not possible to operate and maintain the machine with the

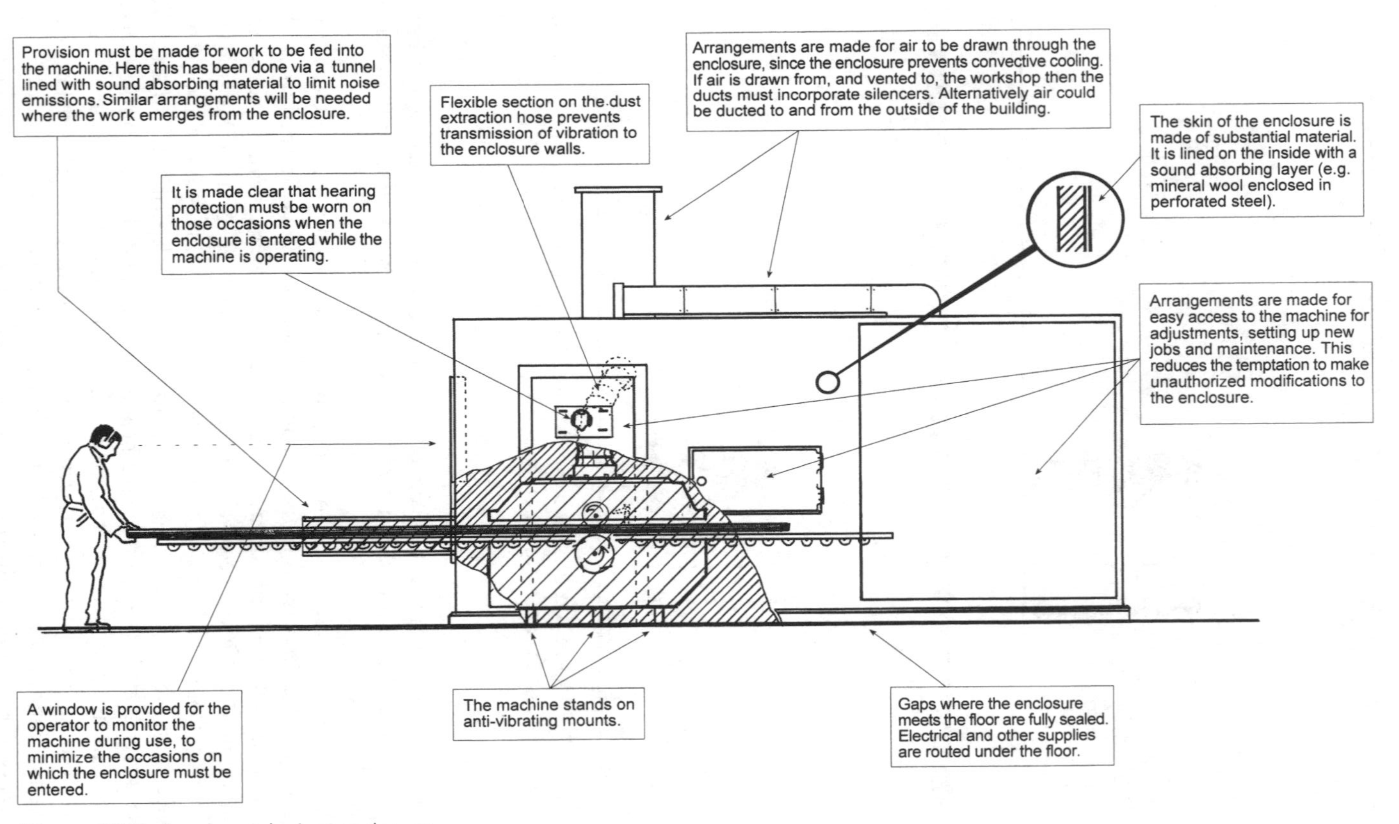

Figure 16.8 A noise reducing enclosure.

noise enclosure in place then it is more likely that it will be partly or completely removed or modified. Different access arrangements may be needed for in-service adjustments at one extreme and for major overhauls at the other. It should be remembered that noise levels inside the enclosure will be much higher than the levels near the machine before it was enclosed. Normally hearing protection will be required inside while the machine is operating, and automatic closing mechanisms and careful control of access can help to limit the periods for which access doors and covers are left open.

Sound absorbing treatments on the inside of the enclosure will be less vulnerable to physical abrasion than in an open factory. They will still need to be chosen with an eye to any contamination by oil and other materials, as well as the possibility of damage during maintenance work.

As well as supporting the machine itself on carefully chosen antivibration mounts, any services which breach the enclosure will probably need to be isolated. Flexible sections are available for most sorts of pipe, duct and conduit, and should be inserted between the machine and the enclosure since the large panels of which most enclosures are constructed can act as effective radiators of sound – particularly at low frequencies – if vibrational energy is allowed to reach them. The rigid pipe or duct sections may need extra support once the flexible section is inserted, and it is important that any additional support arrangements do not bridge the resilient section.

Case study 16.1 Prioritizing noise control measures

A circular saw and its dust extraction system are the main noise sources in a department manufacturing packing cases. The controls are interlocked so that it is possible to operate the extraction on its own, or both machines together. The saw cannot be operated on its own.

'A' weighted sound pressure levels at the nearby assembly area were 92 dB with both machines in operation, and 87 dB when the extractor operated on its own. Two options had been identified for reducing noise levels. The extraction system could be moved out of the department, in which case the noise from the saw alone would remain. Alternatively, the saw manufacturers offered a 'hush kit' which was claimed to reduce noise emissions by 5 dB. Costs in each case would be similar.

In the absence of direct measurements, it was necessary to predict the sound pressure level due to the saw operating without a contribution from the extraction system. This can be done using equation 1.9:

$$L_2 = 10 \times \log(10^{\frac{L_p}{10}} - 10^{\frac{L_1}{10}}) = 10 \times \log(10^{\frac{92}{10}} - 10^{\frac{87}{10}}) = 90.3 \approx 90\,\text{dB}$$

This is also the predicted level with the extraction system moved outside. With the hush kit fitted, the saw alone would result in a sound pressure level of 90 – 5 = 85 dB. Combined with the extraction system, this would result in an L_p of:

$$L_p = 10 \times \log(10^{\frac{85}{10}} + 10^{\frac{87}{10}}) = 89.1 \approx 89\,\text{dB}$$

The conclusion, therefore, is that fitting the hush kit would reduce the sound pressure level in the assembly area to 89 dB, whereas moving the extraction system outside would reduce it only to 90 dB. Other things being equal, the hush kit would be a better noise control measure. Using both methods together could result in a very useful reduction from 92 to 85 dB.

17

Hearing protection

Types of hearing protection

On the face of it, the duty to supply hearing protectors to those employees who require them should be a straightforward matter. Suppliers of personal protective equipment (PPE) normally list many different types in their catalogues. Prices can be as low as a few pence for a pair of ear plugs, and delivery times are short. Yet it is not necessary to visit very many workplaces in order to find examples of establishments where hearing protection is managed very badly. Situations such as:

- The building site where several notices call for hearing protection to be used, but none is available.
- The factory where hearing protection is readily available, but most workers do not seem to be using it.
- The printing works where everyone has some hearing protection on or about their person, but few are using it in the correct manner.

In situations such as these, money is being wasted. Worse, managers are failing to protect their employees' health despite having a legal duty to do so, and despite having identified a method – the use of hearing protection – by which this can be done.

The wide range of different hearing protectors available is an indication that it is not such a simple matter as might appear at first. The price of hearing protection can range from a few pence to over a hundred pounds for a pair. Suppliers do not manufacture or stock equipment for which there is no demand, so each type of hearing protection must be suitable for particular circumstances. It may be tempting to buy the cheapest available hearing protectors, but this will not necessarily be the best choice.

The various types available are divided here into three groups on the basis of their physical construction:

- Ear muffs
- Ear plugs
- Canal caps or semi-aural protectors.

Although these are all forms of hearing protection, the term 'hearing protector' is sometimes applied specifically to ear muffs.

The physical properties of hearing protectors are covered by the various parts of the European standard EN 352 (published as BS EN 352). The most important parts of this standard are Part 1, which covers ear muffs, and Part 2 which covers ear plugs and semi-aural protectors. In each case, the standard prescribes the physical size of the protectors and the way in which it is to be measured. In the case of ear muffs and semi-aural protectors, the range of forces to be exerted by the headband is related to the physical dimensions of the head to which they are fitted.

Ear muffs

Ear muffs, or circumaural ear defenders, consist of a flexible headband carrying a pair of plastic shells (Figure 17.1). The edges of these shells are cushioned,

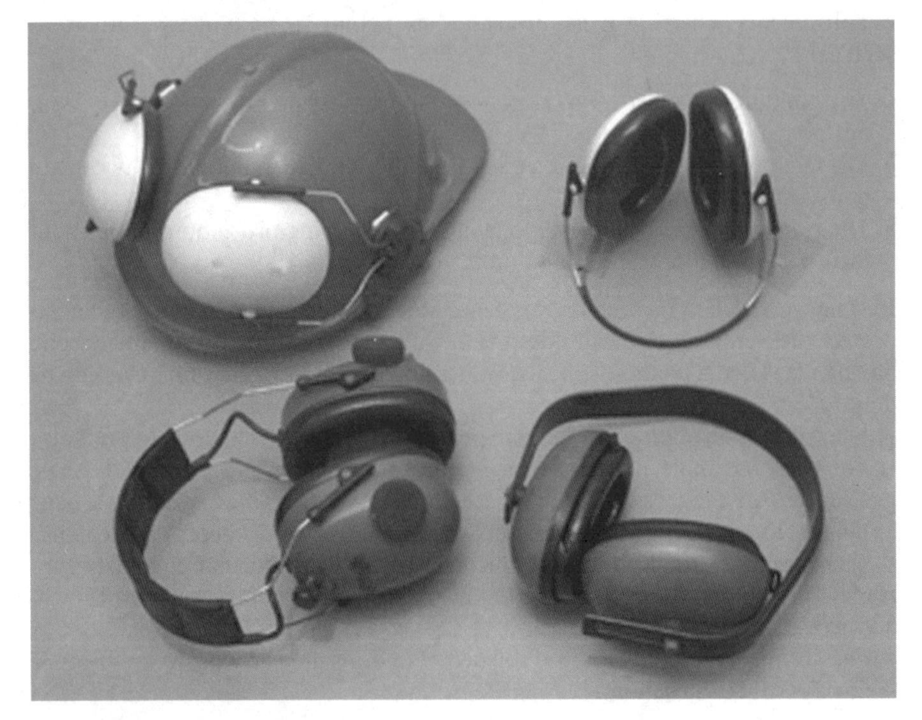

Figure 17.1 Different types of ear muff: helmet mounted, neck band, active muffs with built-in microphones and traditional head band types.

normally by foam enclosed in a thin plastic cover (although liquid filled cushions have also been used). Inside the shell is a foam pad which acts to prevent the reflection of sound inside the shell. The position of the shells on the headband is adjustable to allow for the normal range of head sizes. The effectiveness of this type of hearing protector depends on a tight seal being formed between the cushion and the side of the head. This effectiveness will be reduced if damage or deterioration of the cushions occurs. After a while the plastic covering of the cushion tends to harden so that it is less able to mould itself to the shape of the wearer's head. This hardening takes place much faster in hot conditions. The effectiveness of the seal between the earmuff's cushion and the wearer's head is also affected by objects inserted between them. These objects can include hair, jewellery, thick spectacles or other protective equipment, and clothing. The effectiveness of the seal will also be compromised if the headband exerts a reduced force either because of accidental damage or because the user has deliberately reduced the force to improve comfort.

The standard ear muffs described above are made in cheap, basic versions or in more expensive forms. Some of the more expensive muffs have heavier duty cups and more effective cushions to improve the sound reduction achieved. Some, though, may be designed for improved wearer comfort and this is desirable in itself as employees are more likely to wear them consistently if this can be done comfortably.

Sometimes the standard headband which fits across the top of the head is not appropriate. This is normally because of mutual interference between it and other clothing and protective equipment. Some ear muffs are mounted on slimmer bands intended to fit under headgear such as a hard hat, but they are still not particularly easy to wear in this way. Muffs can be bought which have a neck-band instead. This is a springy wire band which can be worn either under the chin or across the back of the neck. Instead of using either type of band, the cups can be attached via sprung mounts on either side of a hard hat. It should not be assumed that similar ear muffs using either of these methods will provide the same protection as those with a normal headband. The protection of the neck-band type may even depend on whether it is worn in front or behind the head. Suppliers should be able to provide separate data on the protection afforded using each type of support.

More sophisticated – and considerably more expensive – ear muffs incorporate communication or entertainment systems. 'Tactical' ear muffs contain a microphone, amplifier and loudspeaker, which are designed so that the amplified sound remains well short of 85 dB. These are particularly useful for employees who are exposed to noise intermittently but who also need to communicate regularly with others. They are, for example, particularly popular on firing ranges since they allow instructors to be heard clearly without the risk that shots are fired before everyone present has replaced their hearing protection.

Fitting ear muffs is relatively straightforward. The user needs to be aware that the size can be adjusted – and how to do so – and of the importance of excluding objects which would otherwise break the seal. They need to be informed of the need to replace or repair them when damaged, and to watch out for hardening of the cushion. One big advantage of ear muffs as opposed to other protectors is that it is extremely easy for supervisors and managers to observe that they are being used correctly.

Although cheap ear muffs are best disposed of when they deteriorate, hygiene kits are normally available for the more expensive models, including those with the various communication systems built in. These consist of a pair of replacement cushions and absorbent pads. As well as allowing refurbishment, they make it possible for muffs to be transferred hygienically between employees.

Ear plugs

Nightshift workers and those with noisy neighbours are still known to resort to cotton wool in their ears, even though it has been known since the days of Thomas Barr in the 1890s that its physical characteristics do not make it very effective for this purpose.

More recently, it was possible to buy soft glass wool designed to be rolled up and inserted in the ears. In some factories the recommended quantity could be obtained by pulling a lever on a dispenser. This type of 'roll-your-own' ear plug is no longer available. The protection obtained was not very predictable, and there was also the possibility that fibres would become detached and cause problems in the ear canal. The modern version of this type of ear plug uses a preformed plug of glass wool, enclosed in a thin plastic covering (Figure 17.2). This type of ear plug is very cheap, but it has only a limited ability to mould itself to the ear canal in which it is inserted. Because the diameter of the ear canal differs between different individuals, it is necessary to manufacture this type of plug in two or three different sizes to fit different ears. An experienced person – frequently an occupational health nurse – is required to advise each individual which size to use. As with any ear plug, a qualified person is also required to show how to fit it correctly.

Foam ear plugs have the advantage that 'one size fits all'. They are made of a foam material which recovers its original size only slowly after being compressed. To fit them, the wearer rolls them to a small diameter and inserts them in the ear canal. They then gradually expand to fill the space available. They can be reused a few times, but they have one major inherent problem: in the course of rolling them to their reduced diameter any dirt, grease, metal swarf, etc., on the wearer's hands is transferred to the plug and thence to the ear canal. If they are reused, this increases the likelihood that foreign matter will enter the ear and cause or exacerbate an ear infection.

A variation on the foam plug uses a smaller quantity of foam attached to a plastic stem to make a mushroom-shaped ear plug. The stem is used to push the foam, which does not need to be touched, into the ear canal so that the contamination problem is much reduced. In the (admittedly unlikely) event of the stem

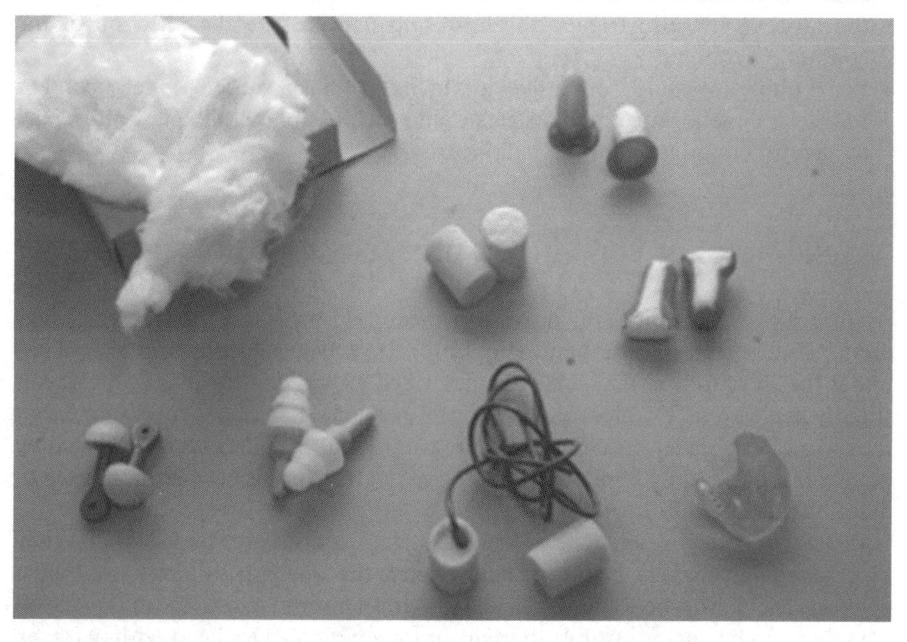

Figure 17.2 Different types of ear plug: top, glass fibre (no longer available), glass fibre in plastic sheath. Middle, two types of foam ear plug. Bottom, foam on plastic stem, soft plastic with flanges, banded foam plugs, individually moulded.

becoming detached, though, medical intervention is required to remove the foam component from the ear.

A more expensive type of ear plug consists of a stem with a number of soft plastic flanges. The flanges are designed so that they fill the ear canal, and the plug as a whole can easily be washed as – in view of the cost of this type of plug – they are normally expected to last for a considerable time.

Even more expensive are the ear plugs which are moulded to fit a particular wearer's ears. It is normally impossible to insert these plugs incorrectly, so that it is easier than with any other type of ear plug for managers to be sure that they are effective at reducing the noise exposure of the wearer to the required level. However, the cost and the fact that a cast needs to be made of each employee's ears, limit the range of situations in which they are used.

In the food processing and precision engineering industries, there is some concern about the possibility of lost ear plugs finding their way into the product. Various modifications can be applied to each of the types of ear plugs described to minimize this risk:

- Plugs can be supplied in pairs joined by a cord which is worn behind the user's neck.
- A ball bearing can be moulded into the body of each plug. Many food production lines incorporate metal detectors in case small machine components are carried into the product. Metal-containing ear plugs which fall into

the process will also trigger an alarm and hence cause the rejection of the contaminated item.

- Some plugs are made from a plastic which itself has a sufficient metal content to trigger a metal detector. This gets round the possibility that a few plugs may be manufactured without the ball bearing.

Semi-aural or semi-insert protectors

At first glance, a semi-aural hearing protector looks like a pair of ear plugs attached to a rigid plastic band (Figure 17.3), which in use is normally worn under the chin. Indeed they are often described as 'banded ear plugs', but this is slightly misleading. On closer inspection, it becomes clear from their shape that they work in a slightly different way from traditional ear plugs, as is suggested by the various other names under which they go; 'canal caps' is another term used to describe this type of hearing protector.

Instead of fitting inside the ear canal, the plug sections are shaped so that they block the ear canal by being pressed against the opening by the band which connects them. This pressure is crucial to their operation, and it is also the reason why they can be uncomfortable to wear for long periods. On the other hand, they

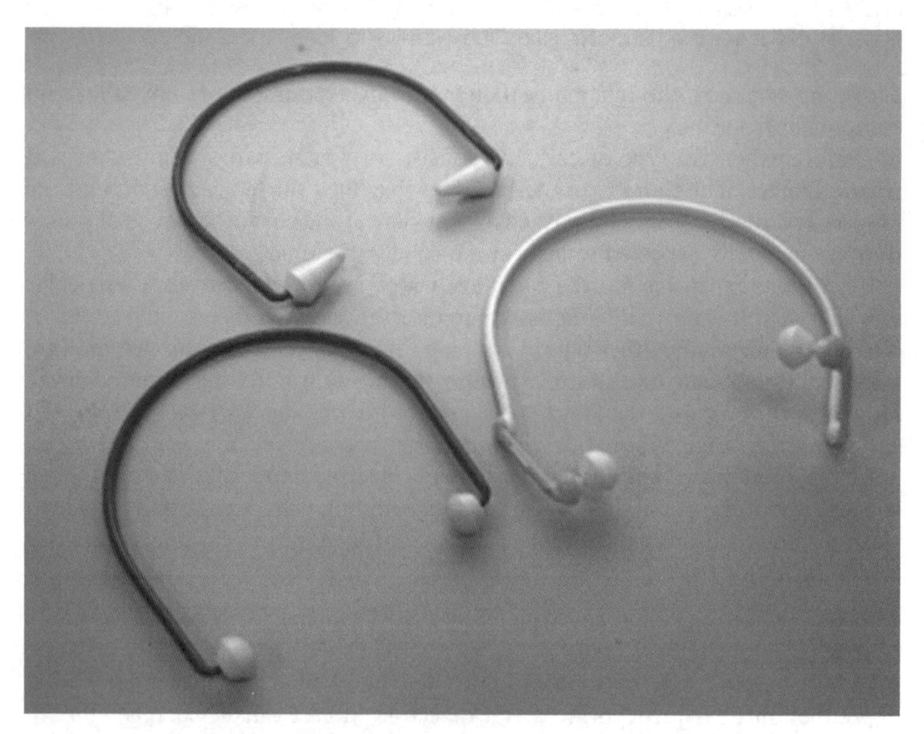

Figure 17.3 Semi-aural ear protectors.

are small and light and it is relatively easy to check that they are being used. They can easily be carried in a pocket – or worn round the neck – by those who are likely to visit a hearing protection zone in the course of a day most of which is spent in quieter areas.

A range of different shapes can be bought and each wearer may find a different type more comfortable. Some have folding bands to make them even easier to carry in a pocket.

The acoustic performance of hearing protectors

The purpose of hearing protection is to reduce the noise exposure of the wearer to a level at which the risk of hearing damage is acceptable. Specifically, in terms of the Noise at Work Regulations, its function is to ensure that the $L_{AEP,d}$ to which the wearer's ear is exposed is below the second action level of 90 dB. Under the regulations which eventually implement the Physical Agents (Noise) Directive, the requirement will be to reduce this exposure below the exposure limit value of 87 dB. As far as the peak action level of the NAWR (and the peak action value of the new directive) are concerned, there is an additional objective of reducing the value of L_{peak} below 140 dB in situations where it would otherwise have exceeded this value.

In order to assess the value of $L_{AEP,d}$ at the ear of the hearing protection wearer, we need to know the sound pressure level at the ear and also the duration of the exposure. Different hearing protectors will reduce the sound pressure level by different amounts, but the effect of any hearing protector will vary with frequency; normally there is a tendency for the protection achieved to increase with frequency. This makes it rather difficult to express simply the effectiveness of any hearing protector. Ideally we would like to know by how many decibels the A weighted sound pressure level is reduced at the ear of the wearer as compared with the sound pressure level when no hearing protection is worn. However, this reduction depends not just on the type of hearing protector, but also on the type of noise – and specifically the sound pressure level in each frequency band – to which it is exposed.

To add to the complications, it seems that different wearers will be afforded different degrees of protection by the same hearing protector. This is true even when it is ensured that in each case the hearing protection is being worn correctly, and it is caused mainly by the differences in the shape and size of the skull and the ear canal between different individuals. Any method for assessing the effectiveness of hearing protectors must take account of the different performance at different frequencies as well as of the difference between different wearers. It must be sufficiently simple to be used in practice and it must be sufficiently accurate to result in effective protection of the hearing of all those affected. There are a number of ways of calculating the effectiveness of hearing protectors. They vary mainly in their degree of complication and in the likely accuracy of their predictions, with the simpler methods normally resulting in a potentially greater discrepancy between predictions and actual performance. However, they all take as their starting point the data provided by the manufacturer or distributor of the protectors, and which is the result of laboratory measurements of the attenuation achieved at different frequencies.

ISO 4869 is the standard which is used when measuring and assessing the effectiveness of hearing protectors. Part 1 describes the test procedure to be used, while Part 2 deals with the various calculation methods – described later – to be used in order to move from the measured attenuation figures to a prediction of the hearing protector's effectiveness in a particular noise environment.

The test used is a subjective one – that is to say it depends on the responses of a panel of subjects (objective test procedures also exist, but they are used mainly for checking the consistency of protection provided by batches of hearing protector and are not a reliable guide to the level of protection afforded to human wearers). In the test, a panel of subjects is selected, each of whom has been shown to have normal hearing. The panel members are first trained in the test procedure. Each subject then sits in an anechoic room surrounded by an array of loudspeakers designed to minimize the effect of slight head movements. A series of pulses of noise is generated containing a narrow band of frequencies centred on the frequency of interest. These pulses are radiated by an array of loudspeakers, and the subject presses a button to indicate if the sound has been heard. In many ways, the test is very similar to the procedure for testing an individual's hearing ability which was described in Chapter 2, the main difference being the type of noise used and the fact that it comes from loudspeakers rather than headphones. As in the hearing test, the hearing threshold is measured at a number of different frequencies, but this time the apparent hearing threshold is measured both with and without hearing protection. The protection provided by the hearing protector at any frequency is the difference between the hearing thresholds with and without hearing protection. The frequencies used are the centre frequencies of the octave bands from 63 to 8 kHz.

The values of the measured protection provided to each individual are collated. For each frequency the mean (i.e. the average) protection received by the different panel members can be calculated, as can the standard deviation. The standard deviation is a value used by statisticians to express the amount of spread in a set of measured values. If the results are close together the standard deviation will be small, but if they are more spread out the standard deviation will be greater. It is a property of many quantities which are distributed in a random manner about a central value that two-thirds of the total number will be found within one standard deviation of the mean. In concrete terms, if at a particular frequency the test results for a hearing protector have a mean of 26 dB and a standard deviation of 4 dB, then it can be assumed that if a large number of subjects are tested, two-thirds of all the results will be in the range from 22 to 30 dB (Figure 17.4).

Another property of this type of distribution of results is that in the case of a few individuals, the measurement will come up with figures that are either very much greater or very much lower than the mean.

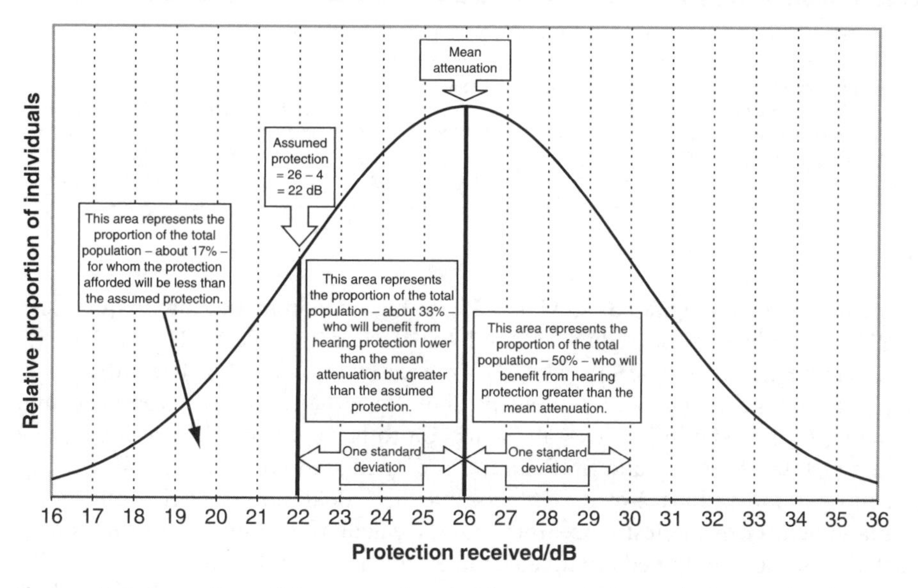

Figure 17.4 Proportions of the population deriving a particular level of protection from a hearing protector with a mean attenuation of 26 dB and a standard deviation of 4 dB.

It is not feasible to measure the protection provided to each individual who needs hearing protection and to come up with an individual solution. The time taken would be far too great even if the equipment and skilled individuals to do the measurements were available in every workplace.

If hearing protection was specified on the basis of the level of hearing protection received by the least protected individual, this could mean specifying a type of hearing protection that was expensive and unnecessary for most of those involved. On the other hand, to use the average protection received as a basis for choosing an effective type would be rash. Half the population will get less protection than this and some will get substantially less. The compromise used is to assume that at each frequency the protection afforded is one standard deviation less than the mean; this value is known as the *assumed protection* of that hearing protector at that frequency, and is easily calculated once the mean and standard deviation have been measured.

Suppliers of hearing protectors make available the mean and standard deviation for the sound level reduction as measured in laboratory tests. Frequently they will also quote the assumed protection, but this is easily calculated as long as the mean and standard deviation in each frequency band is declared. Ear muffs normally carry this information on their packaging. Individual packs of ear plugs are not normally sufficiently large for all this information to be printed in a readable form, but it will be available in manufacturers' data sheets and suppliers' catalogues (Table 17.1).

Also available on packaging and in catalogues will be a few other quantities which are derived from the octave band attenuation figures. These quantities

Table 17.1 An example of the data supplied by hearing protector manufacturer

Frequency/Hz	63	125	250	500	1k	2k	4k	8k
Mean attenuation/dB	14.0	13.0	11.8	18.8	28.9	29.9	37.2	31.9
Standard deviation/dB	4.5	3.5	2.5	2.6	2.9	3.2	4.1	6.2
Assumed protection/dB	9.5	9.5	9.3	16.2	26.0	26.7	33.1	25.7
	H	28	M	19	L	13	SNR	23

include figures designated H, M and L (for high, medium and low frequencies), SNR (standard noise reduction) and for devices on sale in North America possibly a quantity described as NRR (noise reduction rating). The latter quantity is not normally used in Europe, but the others are used for particular methods of predicting the overall protection to be expected. The precise manner in which the H, M, L and SNR figures are calculated is rather complicated – it can be found in Part 2 of ISO 4869 – but this calculation is carried out by the manufacturer or the test laboratory and the calculations using these values once they have been obtained are much simpler and are described below.

Three methods are described here for predicting the actual sound level to which an ear is exposed after the effect of the hearing protectors has been taken into account. In each case information from the manufacturers or distributors is combined with measurements at the particular workstation involved.

Assessing the effectiveness of hearing protectors – octave band calculations

The octave band method of assessing the effectiveness of hearing protection in a given noise environment is the most rigorous, but also the most time-consuming method available. Apart from the data supplied by the manufacturer of the hearing protection, it requires measurements of the sound pressure level in each octave band at the wearer's working position. This requires the use of a sound level meter with octave band filters. Although there are a great many of these on the market – some which measure the different bands simultaneously and others with which the bands must be measured in turn – they are more expensive than sound level meters without these filters and a great many workplaces which can justify the expense of a standard sound level meter will not have octave band equipment available. It is because of this that the other methods described in the next few pages have been developed.

The procedure for predicting the A weighted sound pressure level to which the protected ear is exposed is summarized in the steps below and illustrated in the example which follows:

1. Measure the sound pressure level at the working position in octave bands from 63 to 8 kHz.

2. From the hearing protector supplier's data, calculate the assumed protection at each frequency (if this is not given explicitly).
3. Subtract the assumed protection from the octave band sound pressure level.
4. Add the relevant A weighting at each frequency.
5. Calculate the overall A weighted level by combining the contributions in each octave band using the method for combining decibels described in Chapter 1.

Apart from the last step, the only mathematical operations involved are addition and subtraction, but it is nevertheless not difficult to go astray in making this calculation if the calculation is not set out systematically. It is likely that anyone needing to do this calculation a number of times will set up a spreadsheet to make things easier. There are a few sound level meters on the market which are loaded with a database containing the test data on a number of common hearing protector types. These meters can combine measurement results with hearing protector data to make direct predictions of A weighted protected levels. However, no such database can be complete and the hearing protection on the market changes frequently so regular updating of the database is necessary.

Assessing the effectiveness of hearing protectors – the HML method

The HML method uses the three values H, M and L (for high, medium and low frequencies) provided by hearing protector suppliers along with the octave band attenuation data. It is also necessary to know the A weighted and the C weighted sound pressure levels at the workstation. The steps in the calculation are as follows:

1. Subtract the A weighted sound pressure level from the C weighted value.
2. Calculate the predicted noise reduction (PNR) as follows:

If $L_C - L_A$ is less than or equal to 2 dB, use the equation

$$\text{PNR} = M - \frac{(H - M)}{4} \times (L_C - L_A - 2)$$

If $L_C - L_A$ is greater than 2 dB, use the equation

$$\text{PNR} = M - \frac{(M - L)}{8} \times (L_C - L_A - 2)$$

3. Subtract the PNR from the measured A weighted level to obtain the predicted A weighted protected level.

Although only A and C weighted levels are involved in this calculation, it is still not difficult to go wrong in carrying it out.

Assessing the effectiveness of hearing protectors – the SNR method

The SNR is by far the simplest prediction method, and it uses the SNR value frequently provided with their products by hearing protection suppliers. The only measurement that is needed is of the C weighted sound pressure level at the working position. The SNR value is subtracted from the C weighted level to obtain a prediction of the A weighted protected level.

Comparing the three calculation methods

It should be clear to anyone reading the above paragraphs that the SNR method is by far the simplest method for predicting the performance of hearing protection. The octave band method, because it explicitly takes into account their performance in different frequency bands, is assumed to be the method which makes the most accurate predictions. It is by comparison with the results of this method that the accuracy of the other methods are assessed. They are both much simpler than the octave band method. Under what circumstances will they be adequate?

The spread of results using the three methods can vary up to about ±3 dB; it tends to be greater if either the sound pressure level in one frequency band is much greater than in the others, or if very low or very high frequencies dominate. It is claimed that the HML method is more accurate than the SNR, but it may be doubted whether any improved accuracy justifies the extra complications of the method.

One possible approach, which reduces the volume of calculation required, is to use the SNR method routinely to screen the various hearing protectors available. A shortlist can be produced and the octave band method can then be used to check that the hearing protectors on the shortlist do indeed meet the requirements of the particular situation where they are to be used. In many cases, a simple SNR calculation will show that the protection provided is considerably greater than is required, in which case octave band calculations are not required. Case study 17.1 illustrates the three calculation methods.

The effectiveness of hearing protectors in practice

The test method described in ISO 4869 Part 1 is used to generate manufacturers' declared values used in all three of the methods for estimating the effectiveness of hearing protection. It is carried out under laboratory conditions which in several respects may not be representative of the actual conditions under which the hearing protection is used. For example,

- The hearing protectors under test are brand new ones. The performance of hearing protectors will gradually deteriorate with use.
- The subjects sit still throughout the test. In a real workplace they will be moving about.
- It is ensured that the hearing protection is correctly fitted before and during the test. In a real work situation hearing protection is not always worn properly.
- The figures generated are of the protection received while the hearing protection is in place. It is not possible to take account of periods during which it may be removed.

As a result of these factors the protection afforded by hearing protectors will be less than that which is calculated using any of the available methods. In some cases it will be significantly less than that calculated. Although the assumed protection values used in the calculation methods are less than the mean attenuation measured, it is important to recognize that this difference is incorporated to allow for the differences in protection received by different individuals. It is not intended to include an allowance for the shortfall in protection received due to the factors listed above (Figures 17.5 and 17.6).

Figure 17.7 relates to a specific situation where the unprotected sound pressure level is 100 dB and the protected level at the ear (as assessed using the procedures described earlier) is 70 dB. The graph shows that in this particular situation, the exposure could reach the lower exposure action value of the Physical Agents (Noise) Directive if the protection was removed for as little as 1 per cent of the shift: about 5 min. The second action level of the 1989 regulations is reached if the hearing protection is removed for 10 per cent of the shift.

The UK Health and Safety Executive (1998) has recommended that an allowance should be made when calculating the effectiveness of hearing protectors for the factors discussed above. They suggest that the assumed protection be reduced by 5 dB for ear muffs and by as much as 18 dB for ear plugs (the latter figure raised some eyebrows when it was first published). This is probably unsatisfactory advice. Although a small allowance is appropriate for unavoidable reduction in performance in real work situations, it would be preferable to minimize the effects of these factors by good management of hearing protection. In the absence of good management, the figures suggested by HSE may be insufficient.

Another reason why it is not desirable to make arbitrary adjustments to the predicted effect of hearing protection is that if, despite this allowance being made, hearing protection is used properly, employees may be overprotected against noise. This has the undesirable effect of preventing communication and isolating the wearer from useful information, such as the sound of a vehicle moving nearby or a warning alarm. It may make it less likely that the employee will use hearing protection consistently and thus have the paradoxical effect of increasing noise exposure. In most situations a target daily equivalent protected level of between 70 and 75 dB is probably ideal.

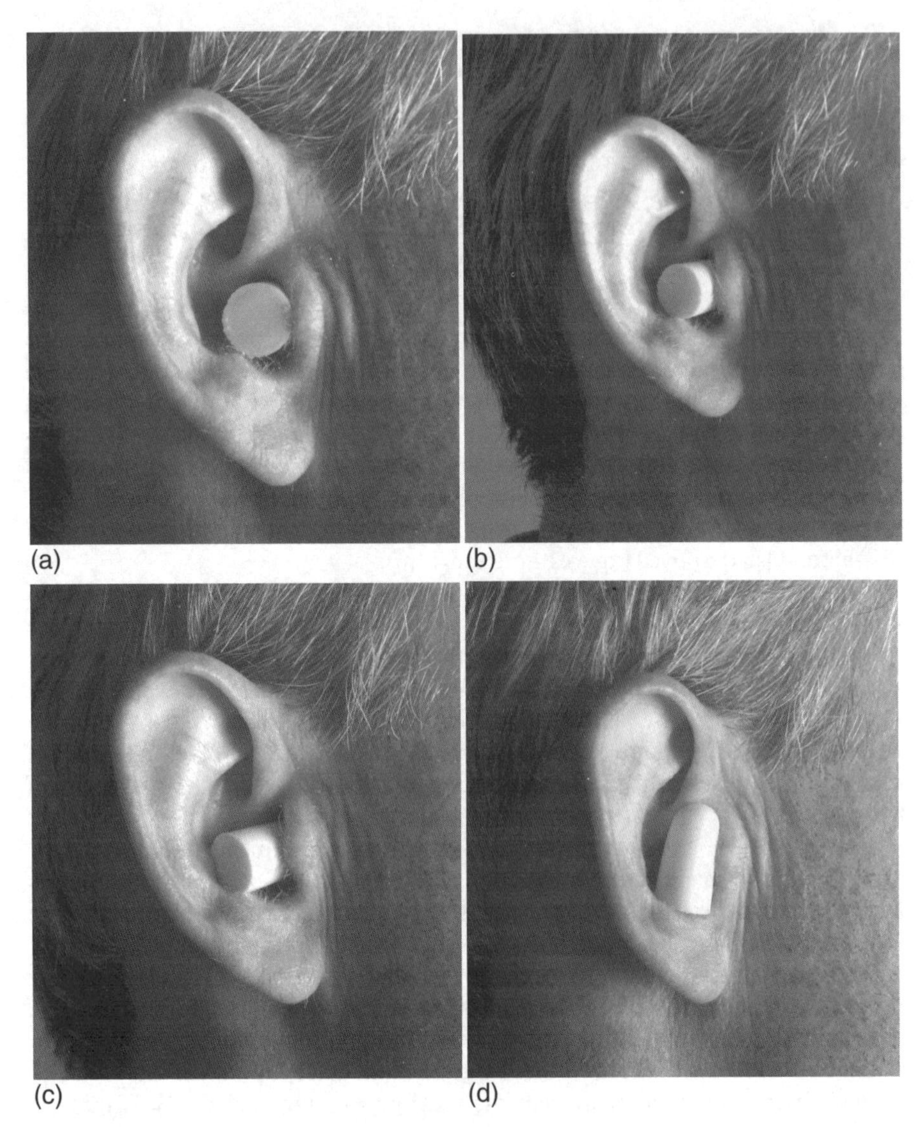

Figure 17.5 Ways of wearing an ear plug: (a) correct; (b) not fully inserted; (c) wedged in the end of the ear canal; (d) not in the ear canal at all.

Inspecting and maintaining hearing protectors

It is a duty of employers to maintain hearing protection provided under the Noise at Work Regulations. In many workplaces, the condition of all personal protective equipment (PPE) will be regularly monitored while in others hearing protection may need to be the subject of a special programme of maintenance.

Figure 17.6 Badly fitted ear muffs with their effectiveness reduced by (a) hair

(Continued)

Where disposable ear plugs are used, the maintenance of ear protection can be reduced to a simple matter of checking regularly that dispensers are stocked, but 'regularly' in this context is likely to mean several times a week. As back-up, a notice saying where further supplies can be obtained can guard against unavailability due to unexpectedly high use.

If reusable ear plugs or ear muffs are in use then inspections do not need to be as frequent as this, but a more comprehensive set of checks are required. The checks made, and any action taken as a result, should be recorded. Suitable storage facilities must be present. These may have the function of making hearing protection available at points where it is required, or of protecting it from dust and other

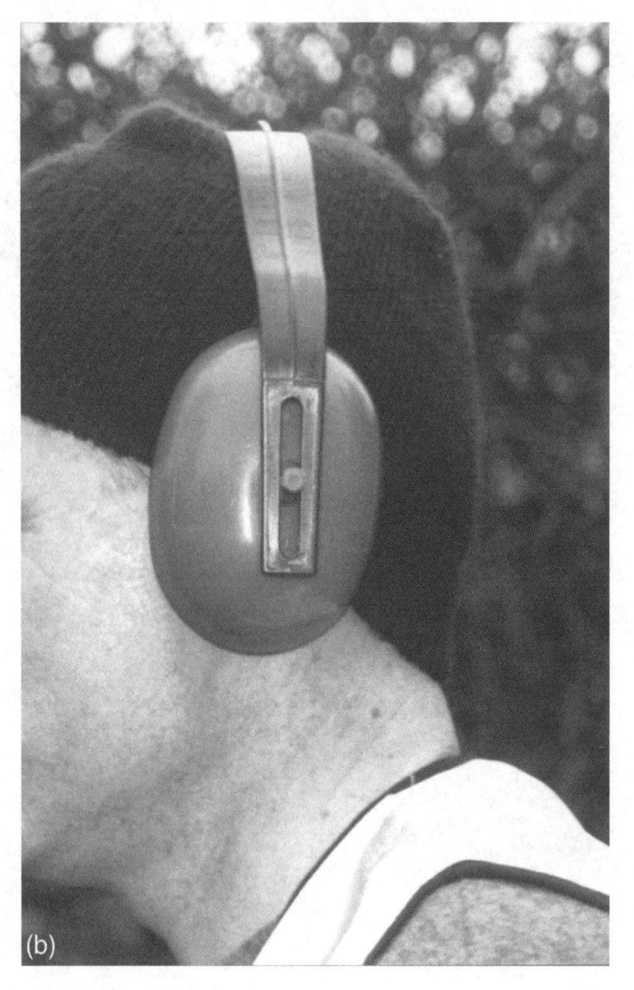

Figure 17.6 (continued) and (b) clothing.

harmful agents. If hearing protectors are intended for use by several individuals, then cleaning/disinfecting facilities are needed at the storage point.

Individual hearing protectors should be examined for signs of deterioration, accidental damage or unauthorized modification. A common form of deterioration is a hardening of the plastic cover round the foam cushions. This can take place very quickly under hot conditions. This cushion can be accidentally torn, and the sound absorbing pads inside the shells can be lost or removed or can become clogged with grease or dust. For expensive ear muffs, the hygiene kits described earlier in this chapter can be used to replace worn or damaged components (Figure 17.8). Deliberate damage to ear muffs can include the reduction of the force exerted by the headband, or the drilling of holes in the cups. Replacement of the muffs is normally necessary if this kind of damage has occurred.

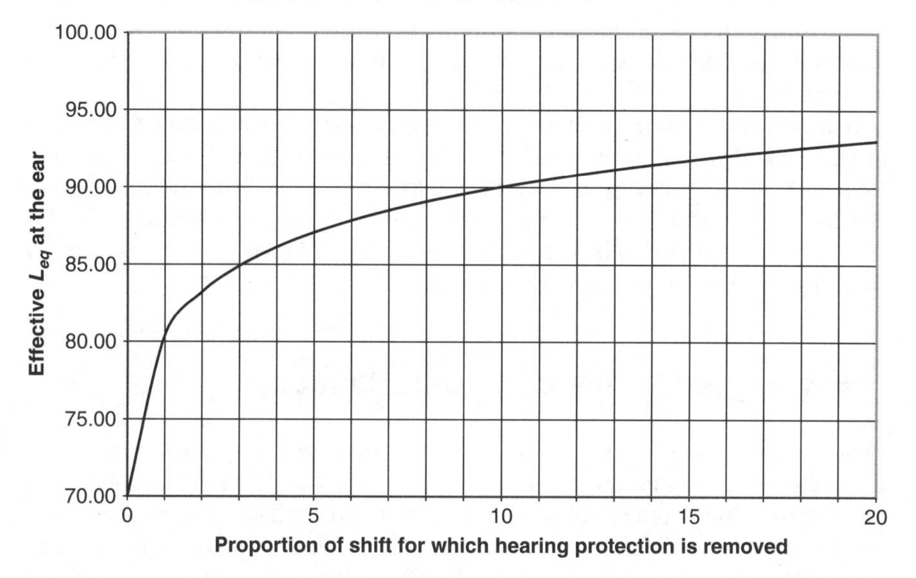

Figure 17.7 An illustration of the effect of removing hearing protection for part of the shift.

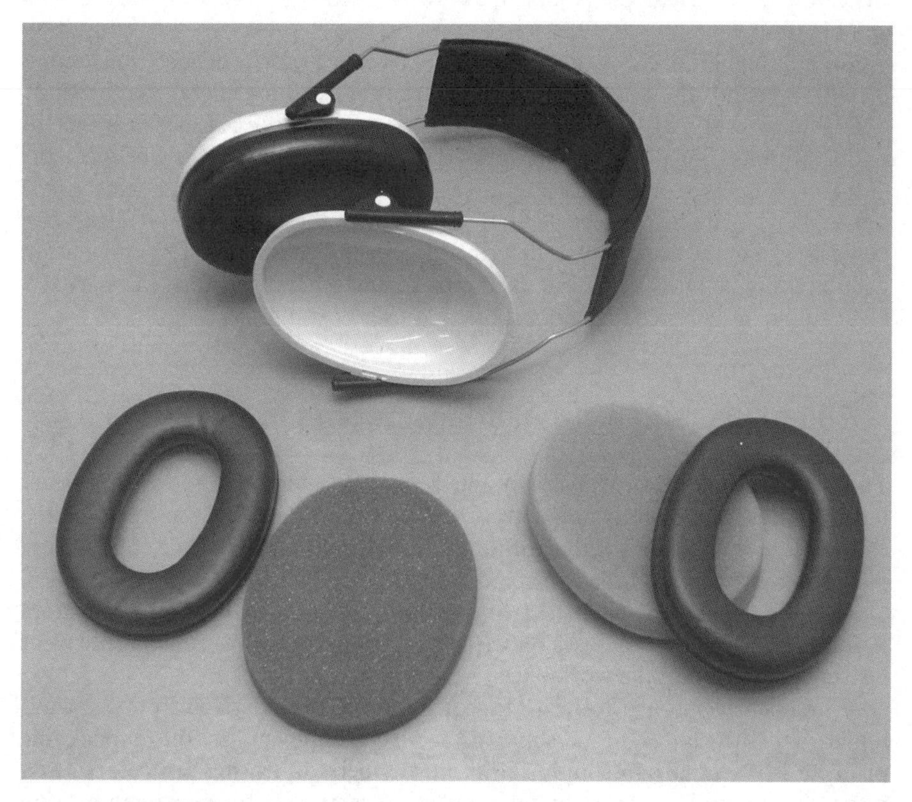

Figure 17.8 Replacement of cushions and foam pads.

Hearing protectors stored at the working position can be inspected when not in use. Muffs or reusable plugs kept by the employee will need to be produced for inspection.

If a number of repairs and/or replacements are needed during a round of inspections, then this is evidence that the inspections should take place more frequently. It is of course a duty of employees to report any defects in their hearing protection and it should be made clear how they are expected to do this. It may be appropriate to include in the inspection of hearing protectors a check that this information is prominently displayed.

Hearing protection in special situations

In some situations the types of hearing protection described above may not be adequate or may not be appropriate. To protect against very high noise levels it is possible to wear ear plugs and muffs together. The overall protection is limited by the transfer of sound energy to the inner ear via the skull, so that it is not possible to add the protection offered by the muffs to that resulting from the plugs. Instead, the particular plugs and muffs to be used will need to be tested as a combination.

For still higher levels of protection it is possible to use a helmet which completely surrounds the head to prevent this transfer of sound energy via the skull.

In some jobs it may be necessary to wear protective headgear which is incompatible with hearing protection. Examples of this include motorcycle helmets, the anti-riot helmets worn by police forces, or helmets to protect against toxic atmospheres. The hearing protection provided by these types of headgear can be measured in the same way as described earlier in this chapter. The attenuation will normally be lower than that expected from hearing protectors, but may still be adequate in the specific circumstances of use.

The management of hearing protection

This chapter, and earlier parts of the book, have discussed in detail many individual aspects of hearing protection. The purpose of this final part of the chapter is to emphasize the integration of the various aspects of the effective use of hearing protection in the workplace.

Reference has been made to two standards relating to hearing protection. A third – BS EN 458 – advises on various issues to do with the management of hearing protection in the workplace.

A simplistic approach to hearing protection could be derived from the catalogue of a distributor of personal protective equipment. In this model, the problem is essentially to supply sufficient numbers of hearing protectors at an affordable price. This is clearly not adequate as a strategy for managing hearing protection, even though there are many workplaces where this summarizes the extent of planning that is devoted to the subject.

Many managers would see a need to go further than this. Perhaps unfortunately, a great deal more attention has been given in the past to the various ways in which the effectiveness of hearing protection can be estimated – the various calculation procedures detailed in earlier parts of this chapter – than to the selection of suitable protection. There are a great many workplaces where it is vital to carry out the necessary calculations for all or some of the employees affected by noise. There are also a large number of work situations where noise levels are such that it is necessary to provide hearing protection, but where the levels are within a range where any hearing protection which can legitimately be described as such (meaning that it meets the minimum specification for the attenuation of hearing protectors which can be found in BS EN 352) will provide adequate protection. This is normally the case if a reduction of less than 10 dB is required.

Whether or not calculations are required to assess the effectiveness of hearing protection, this is just one step in its effective management. Hearing protection which is able to provide the necessary attenuation will nevertheless not do so if it is not worn for all or most of the period during which noise exposure takes place. The reasons for not wearing hearing protection – as with other sorts of PPE – can be many. Managers traditionally blame employees for not using the protection provided, and this is undoubtedly an important factor. However, there are many reasons which may fall within managers' area of responsibility. Ultimately if hearing protection is being provided and not properly used, then responsible managers will need to find out why this is happening and take steps to ensure its effective use.

Some reasons why workers do not use hearing protection which is provided for their benefit could include:

- The type provided is uncomfortable
- Workers have not been given training and information in the need to protect their hearing
- Dispensers are not topped up regularly, or ear muffs are not inspected and replaced when necessary
- The job is designed in such a way that it is difficult or impossible to carry out with hearing protection in place. This could be because:
  - It is necessary to speak regularly to colleagues.
  - Noise from the machinery gives important information about its operation.
  - Safety information such as the movement of vehicles is lost when using hearing protection.
  - Hearing protection is incompatible with other PPE.
- The boredom of the job may only be relieved by social interactions and entertainment, such as listening to music while working.
- Managers and supervisors do not feel that to ensure the use of hearing protection is their responsibility.

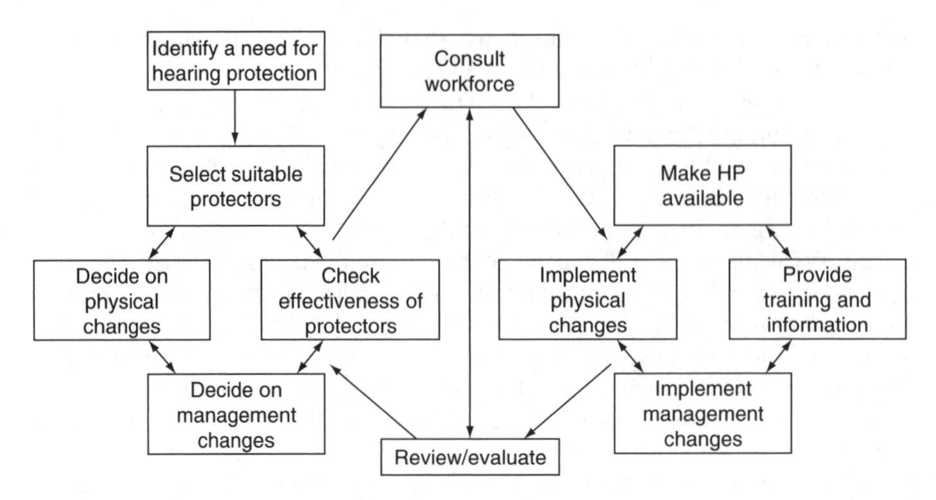

Figure 17.9 The management of hearing protection.

Even where workers attempt to use hearing protection, it is not unusual for it to be ineffective in practice because they do not know how it should be worn, or do not understand the importance of keeping it on all the time when exposed to very high noise levels.

Figure 17.9 shows a model for the management of hearing protection. It is not intended to be definitive. In many workplaces PPE is managed in an integrated way, while for smaller workplaces less sophisticated management practices may be adequate. However, the diagram does attempt to show the most important factors involved in managing hearing protection, and it indicates the ways in which they are related.

The first step is to identify the need for hearing protection. Normally this is the result of an assessment of employee noise exposure and a consideration of various alternative ways of reducing it.

Decisions need to be made about various aspects of the use of hearing protection. The key issue here, and the first one on the diagram, is the choice of suitable hearing protectors. In most cases alternative types will be needed to suit every individual requiring protection. However, decisions about the suitability of hearing protection are crucially connected to an assessment of their effectiveness in the particular work environment concerned, and to decisions about various physical and management changes required in the workplace as a result. The type of physical changes required would include installing ear plug dispensers and storage boxes for muffs; putting up appropriate signage (and possibly removing misleading signs) and painting demarcation lines for hearing protection zones; or replacing audible safety warning systems with flashing beacons so that those using hearing protection are aware of other hazards. Management changes would include training supervisors to enforce hearing protection rules or redesigning jobs so it is no longer necessary to talk regularly to others in a situation where this is most easily done by removing hearing protection.

If the workforce and their representatives are not involved in the decision-making process, then they will at the very least need to be consulted before any changes are made, both in the interests of getting their co-operation, but also to ensure that the decisions made are capable of being implemented in practice. As a result of consultation with the workforce, some companies have been persuaded to spend considerably more than they had originally intended on hearing protection that their employees found comfortable and convenient to use.

There is no need to implement a series of changes in practice overnight, but at the other extreme it is difficult to get workers to take seriously measures that seem to be implemented in a half-hearted way. For example, notices threatening draconian action against those failing to use hearing protectors should not normally be put up before the hearing protection is available. It is essential that workers are shown how to use hearing protection effectively and understand the importance of using it consistently, so training and information sessions should come early in any programme of changes, although they, too, are unlikely to be very effective if they precede the arrival of the hearing protection.

A review of the hearing protection measures should take place after a suitable period, during which any problems will be identified and the original decisions made can be confirmed or altered in the light of experience.

Case study 17.1: Effectiveness of hearing protection

Using the octave band method

The following sound pressure level measurements are made at the operating position of a panel saw:

Frequency/Hz	63	125	250	500	1 k	2 k	4 k	8 k
L_p/dB	87.2	89.4	98.6	98.5	96.4	87.1	75.5	64.8
A weighted	100.8	C weighted		103.1				

The operator works at this position for a full 8-h shift, and wears ear muffs. The manufacturer's data for that particular model involved are as follows:

Frequency/Hz	63	125	250	500	1 k	2 k	4 k	8 k
Mean attenuation/dB	14.0	13.0	11.8	18.8	28.9	29.9	37.2	31.9
Standard deviation/dB	4.5	3.5	2.5	2.6	2.9	3.2	4.1	6.2
	H	28	M	19	L	13	SNR	23

It is necessary to decide whether these ear muffs will provide adequate protection to the operator's ears, assuming they are worn for the whole of the 8-h shift.

1. Calculate the assumed protection at each frequency:

Frequency/Hz	63	125	250	500	1 k	2 k	4 k	8 k
Mean attenuation/dB	14.0	13.0	11.8	18.8	28.9	29.9	37.2	31.9
Standard deviation/dB	4.5	3.5	2.5	2.6	2.9	3.2	4.1	6.2
Assumed protection/dB	9.5	9.5	9.3	16.2	26.0	26.7	33.1	25.7

2. Subtract the assumed protection from the measured octave band sound pressure level to calculate the assumed protected levels (APLs):

Frequency/Hz	63	125	250	500	1 k	2 k	4 k	8 k
Octave band level/dB	87.2	89.4	98.6	98.5	96.4	87.1	75.5	64.8
Assumed protection/dB	9.5	9.5	9.3	16.2	26.0	26.7	33.1	25.7
APL/dB	77.7	79.9	89.3	82.3	70.4	60.4	42.4	39.1

3. Add the relevant A weighting figures to the APL in each octave band:

Frequency/Hz	63	125	250	500	1 k	2 k	4 k	8 k
APL/dB	77.7	79.9	89.3	82.3	70.4	60.4	42.4	39.1
A weighting	−26.2	−16.1	−8.6	−3.2	0	1.2	1.0	−1.1
A weighted APL	51.5	63.8	80.7	79.1	70.4	61.6	43.4	38.0

4. Finally, carry out a decibel addition of the A weighted band levels:

$$L_A = 10^{\frac{51.5}{10}} + 10^{\frac{63.8}{10}} + 10^{\frac{80.7}{10}} + 10^{\frac{79.1}{10}} + 10^{\frac{70.4}{10}} + 10^{\frac{61.6}{10}} + 10^{\frac{43.4}{10}} + 10^{\frac{38.0}{10}} = 83.3\,\mathrm{dB}$$

Using the HML method

Using the data in the previous example, $L_C - L_A = 2.6\,\mathrm{dB}$, so the equation to use is

$$\mathrm{PNR} = M - \frac{(M - L)}{8} \times (L_C - L_A - 2)$$

where $H = 28$, $M = 19$, and $L = 13$.

$$\mathrm{PNR} = 19 - \frac{(19 - 13)}{8} \times (103.1 - 100.8 - 2) = 18.7\,\mathrm{dB}$$

so the protected level will be $100.8 - 18.7 = 82.1$ dB.

Using the SNR method

Using the data in the previous example, C weighted sound pressure level at the working position = 103.6 dB; SNR = 23, so the protected level is 103.1 − 23 = 80.1 dB.

Conclusion

Using the octave band method, the protected level is below 85 dB. This complies with the legal duty under the 1989 regulations to reduce exposure below the second action level, and is below the 87 dB exposure limit value of the Physical Agents (Noise) Directive. There is evidence of a risk to health at this level, and it would presumably be reasonably practicable to provide a higher grade of protector. The target for protection should be in the region of 75 dB. A lower target could be set to make up for the shortfall in the protection actually received, but it is more important to ensure that the use of hearing protection is managed effectively.

The HML and SNR methods produce results within about ±3 dB of the octave band prediction.

Case study 17.2 The choice of suitable hearing protection

Windgate Engineering Ltd services and maintains small commercial vehicles for a number of private companies. Along one side of the modern workshop are a series of bays, each operated by an engineer who carries out a wide variety of operations on vehicles. The other side of the workshop is occupied by specialist departments including the electrical department, the stores and a new but growing department which carries out conversion and customization work. Panel beating and other body work, though, is contracted out to a specialist firm. The manager was appointed recently with the task of establishing up-to-date quality management procedures to improve the company's position when tendering for work from prestigious customers. He has also been working to improve health and safety practices and one of his first acts was to commission a noise exposure survey of the premises. This established that while using several types of tool, the engineers were exposed to A weighted sound pressure levels ranging up to 99 dB. These tools included impact wrenches, grinders and welding sets. The survey noted that the mix of different tasks carried out on different days varied considerably. Assuming a 'typical' day's mixture of different tasks, the personal daily exposure of an engineer due to the use of these tools would be around the first action level, but it was likely that on some days the second action level would be exceeded, and it would be impossible to predict this in advance. When noisy tools were not in use, sound pressure levels at the engineers' working positions were generally in the mid-70s, and even when noisy work was in progress at an adjacent bay it was not observed to rise above 81 dB.

Work in the conversion department is sometimes very noisy. When cutting body panels for the installation of new windows or other fittings, the workers involved were exposed to levels as high as 110 dB over periods of up to 30 min. The engineers in the two bays nearest to this department would also then be exposed to levels above 90 dB, and this would add to the noise exposure resulting from their own work. This work was irregular and difficult to predict, but normally needed to be completed within a short period.

The consultant carrying out the noise exposure survey pointed out some anomalies in the way hearing protection was managed. At regular positions around the workshop walls were signs announcing that hearing protection must be worn at all times. The longer serving engineers all had ear muffs, but these had not been issued to more recent recruits, and some of the muffs in use had seen better days. Those engineers with ear muffs did not use them in any consistent manner beyond the fact that they all ignored the signs on the walls. An ear plug dispenser near the entrance had been empty for some months, even though a supply of foam ear plugs was discovered on an upper shelf in the stores.

It was clear that it was unnecessary to require the use of hearing protection for most of the time. It was uncomfortable to wear for long periods when working on vehicles, and it interfered with speech communications when ordering parts from the stores or when discussing progress with supervisors and managers. However, hearing protection needed to be made available to all employees and they would need to use it at appropriate times. As far as their own work was concerned, it was quite easy to identify those tools and operations for which hearing protection should be compulsory. It is not strictly necessary to protect employees from short exposures above 90 dB, as long as the daily exposure, $L_{AEP,d}$, is below the second action level. By making protection compulsory for even short exposures above 90 dB, though, it can be guaranteed that in situations where exposure periods vary widely from day to day there is no possibility that the second action level will be exceeded. Notices reminding users of the need for hearing protection were attached in each case either to the relevant tool or next to its storage position.

The ear muffs in use were condemned, and it was further decided that the foam ear plugs were not appropriate in an engineering environment due to the risk of introducing oil and swarf into the ear canal. A supplier of personal protective equipment was contacted and a representative visited with a selection of hearing protectors. Samples were left for evaluation and, following discussion with the employees, two models of ear muff were selected between which individual engineers could choose. Storage boxes were installed at each workstation, and arrangements were made for the ear muffs to be inspected and repaired or replaced at regular intervals. A supply of ear plugs which did not need to be compressed before insertion was ordered for use by visitors and by those who preferred them to muffs. Once again, arrangements were made to check the supply of ear plugs and replenish when necessary.

The panel cutting operations in the conversion department were a bigger problem. Various solutions were considered, including making hearing protection compulsory throughout the workshop when panel cutting was in progress. This was thought to be too sweeping. A proposal to sound a siren before starting noisy operations was also rejected as it would add to the noise levels. It was eventually agreed that under normal circumstances panel cutting would take place during overtime when few other employees were on the premises, and that when it was in progress everyone present would be warned to wear hearing protection. On occasion, to meet a deadline, it would be necessary to do the work during normal hours, but panel cutting was only to be carried out with a permit-to-work from the workshop manager or his deputy, who would be responsible for warning those nearby and checking that they used hearing protection. In the conversion department itself, only one model of ear muff, which had been shown by calculation to afford an adequate degree of protection, was permitted.

For the longer term, trials were to be made of ways of reducing noise during panel cutting, including the use of adhesive vibration damping materials.

The signs requiring hearing protection at all times were removed, and a notice was erected in each test bay detailing more precisely the circumstances in which it was required to be used. A notice by the entrance gave details of where new or replacement hearing protection could be obtained if required. A training session was arranged at which all employees could be trained in the effective use of hearing protection, and a follow-up session took care of those who missed the first one.

Reducing hand–arm vibration risks

HAV management strategies

In implementing an HAV management strategy based on a hierarchy of control (see Chapter 15), a clear distinction needs to be drawn between the various types of measure which can be implemented:

1. There are a number of strategies which either reduce vibration levels at the tool handle or which reduce exposure time. Any of these measures will reduce the assessed dose and also reduce the risk of the development of HAVS.
2. There are strategies which limit the transfer of vibration energy from the tool handle to the operator's hand. These strategies will not affect the assessed dose. They will, however, reduce the HAVS risk.
3. A number of measures are thought to reduce the risk of HAVS developing once a given quantum of vibrational energy has been transferred into the hand–arm system. Many of these measures involve maintaining the blood circulation in the fingers. None of these strategies will reduce the assessed hand–arm vibration risk.
4. One apparently useful approach – the use of personal protective equipment – does not in fact reduce either the assessed hand–arm vibration dose or the risk of HAVS.

Reducing vibration exposure

A number of measures are discussed in Chapter 15 which can be expected to reduce the vibration dose as assessed by measuring vibration levels at the handle. Regular maintenance and the selection of low vibration tools are possibly the most useful approaches under this heading and are discussed more fully in that chapter.

The above approach is more likely to be useful in a factory environment than, say, in a quarry or demolition site. Outdoor work lends itself much less to the use of supports and tool hangers. However, some types of outdoor tool can be used with a harness which, if properly adjusted, distributes the weight to the operator's shoulders and allows the use of a much lighter grip (Figure 18.3).

The choice of appropriate, well-designed tools for any particular task is a matter for managers, who will take into account a number of other factors – some of them discussed in Chapter 15 of this book – when making a decision. It is much more likely that operators will use an appropriate technique if they are adequately trained for the job and are consulted about the choice of equipment and the design of the individual task.

Job rotation

If it is decided that a worker is exposed above either the exposure action value or the exposure limit value, then the most immediate way by which this dose can be lowered is to reduce the exposure time by sharing that task among a number of employees. This is in contrast to the situation where a worker is exposed above one of the action levels for noise. In that case the use of hearing protection is the

Figure 18.3 A harness supports the weight of a strimmer.

The use of a resilient material between the source and the operator's hand can, in principle, reduce vibration levels at the hand. The operation of vibration isolating materials and structures is discussed more fully in Chapter 16. The limitation when applying this principle to hand–arm vibration control lies in the fact that the vibration frequencies which are most damaging are relatively low ones. A resilient hand grip, compressed by the operator's hand in normal use, will have a natural frequency which is likely to be rather higher than those frequencies which cause HAVS, and it is unlikely to appreciably reduce the levels of those vibrations which give rise to HAVS. It may well make the tool more comfortable to use, and will protect against high frequency shock components so that a properly designed resilient grip will probably be desirable for other reasons. A DIY attempt to isolate from vibrations – typically a piece of pipe insulation slipped over the tool handle – may increase vibration levels, so this type of unauthorized modification should be avoided.

Some low vibration tools, particularly those used in civil engineering trades, are manufactured with sprung handles, which significantly reduce vibration levels at the handle compared with traditional tools (Figure 18.1). This kind of spring-support system has the low natural frequency which is necessary to achieve a significant reduction in vibration magnitudes across the range of frequencies which are most damaging. It is not normally practical to incorporate this kind of vibration isolation system in smaller, hand-held tools.

Figure 18.1 An antivibration chain saw, showing one of the rubber bushes on which the handle is mounted.

Where low vibration handles are fitted to a multipurpose tool they may be less effective for some operations than for others. For example, some jackhammers incorporate vibration isolation which becomes effective when pressing the tool down against the springs. Operators need to be trained to disengage the drive before lifting the tool, as on the upward stroke the handles are pulled back against rigid stops. For some operations, it is not possible to lift the tool with the drive disengaged, and as a result vibration exposure may be much higher than expected on the basis of published vibration emission levels. This is a situation where measurements made on the job are essential.

Grip and support issues

For a given level of handle vibration, the energy transferred to the hand/arm system, where it can potentially do the damage which results in HAVS, will depend crucially on the tightness of the operator's grip. Not so the vibration level measured at the handle. This may well be unaffected by the tightness of grip. In some cases a tighter grip may actually reduce the vibration levels measured at the tool handle as a result of the extra damping introduced.

If it were practicable to measure the vibration energy transferred to the hand rather than simply the vibration of the handle, this would not be a problem. The measured values would then relate more directly to the probability of damage occurring. Measurements of vibration transferred to the hand–arm system have been attempted, but because of the difficulties involved there is no standardized procedure for doing this, and it is unlikely that one will be developed. Neither is there a way in which tool handle measurements can be adjusted to take account of grip factors.

Whatever the assessed exposure dose, it is therefore desirable to take steps to reduce the transfer of vibration energy from the tool handle to the hand itself. Factors which cause the tool to be gripped more tightly than is strictly necessary include:

- Having to support the weight of heavy tools or work pieces;
- Poor technique;
- Inappropriate tools in use;
- Poorly designed tools.

If a heavy tool has to be lifted and applied to the work, it may be possible to support the weight of the tool on a tool hanger (Figure 18.2), while still allowing it to be manipulated over the range required. If a heavy piece of work is to be applied to a fixed tool, it is often more appropriate to support the work from below. The operator then only has to grip tightly enough to guide the tool or piece of work. Of course, a badly designed support system has the potential to require the operator to use a tighter grip than was necessary before the system was introduced.

Figure 18.2 Use of a tool hanger.

most obvious and most immediate way to reduce exposure, whatever alternative approaches may be available in the longer term.

There are two major problems with job rotation as a strategy for reducing hand–arm vibration exposure. The first is that the total exposure of the workforce is not reduced, but merely spread among more employees. Although individual exposure will be reduced, a greater number of employees will be exposed to levels which may be near the relevant action value, even if they do not exceed it. Since the susceptibility of different individuals varies greatly, it is possible that while keeping within the strict legal requirements, more employees could actually develop HAVS symptoms. The second problem with using job rotation as the main method of exposure control is that it may be necessary to reduce exposure time by a large factor, meaning that in some cases the job would have to be shared between several employees. This situation will occur when existing HAV exposure is very high, and of course in this case it is all the more necessary to reduce the HAV dose. Where options exist for reducing the vibration levels at the tool handle, this is a much more effective way of reducing vibration dose since a halving of vibration levels will have the same effect on the hand–arm vibration dose as a quartering of exposure time. Investing in lower vibration tools and processes may be a more direct way of reducing exposure without requiring major reorganization of work routines.

Maintaining circulation

It is thought that the vibration-induced damage to the circulation and nervous systems of the hand which eventually leads to HAVS will progress faster if the blood supply to the hand is reduced during vibration exposure. The most important factor affecting the blood supply to the hand is its temperature. If it is cold then the body acts to reduce the blood supply to the periphery so as to maintain the temperature of vital organs. This is of course a normal reaction and should not be confused with the extreme effects of cold exposure in sufferers from HAVS and Raynaud's disease. Even with gloves, the circulation in the hand may be affected if the body temperature is low. Many jobs with vibrating tools are carried on outdoors or in similar cold environments and various strategies are available for maintaining temperature:

- The use of tools with heated handles;
- Supply of warm gloves and other clothing;
- Warm-up breaks indoors;
- Hot drinks.

Pneumatic tools are driven by the expansion of compressed air, and air, like any other gas, cools as it expands. The exhaust from these tools can be extremely cold, and there are still a few tools in existence which vent the exhaust air over the operator's hands. Even in a warm building, the hands can as a result get very cold. Other pneumatic tools may vent air away from the operator, but if the tool is used for long periods the handle will itself get very cold. This can be helped if the handle has a thermally insulating exterior, or if the operator wears gloves. In some cases non-air-powered tools would be preferable.

Smoking also has the effect of reducing blood supply to peripheral parts of the body. It is thought likely (the evidence is not conclusive) that smoking during work with vibrating tools, or just before working with vibrating tools, may also accelerate the damage caused by the vibration. It is therefore desirable that users of vibrating tools are discouraged from smoking at work. The effects of nicotine on the circulatory system last for a considerable time, so smoking breaks will be of little benefit if the tool is used again immediately afterwards. It is in any case doubtful whether efforts to persuade existing smokers of the additional risks involved will be very successful in persuading them to give up.

Personal protective equipment

There are a number of antivibration gloves on the market. They normally contain a thick pad between the palm and/or the fingers and the tool handle. The nature of this pad varies. In some cases it consists of a foam material, while in other types it can be a gel-filled sac. Essentially, this resilient material, along with the grip force applied, acts like the simple vibration isolation systems described in Chapter 16. These systems provide useful protection against vibration frequencies well above their natural frequency, while increasing vibration amplitudes at and below this frequency.

The type of system represented by a hand and the resilient pad built into a glove has a natural frequency higher than those mainly implicated in hand–arm vibration syndrome. As a result they do not normally reduce the risk of HAVS significantly. In some cases a small reduction in vibration amplitude is achieved, while in others the vibration exposure of the hand may actually increase. The structure of most antivibration gloves normally makes it much more difficult to grip and control tools and work than would be the case if they were not worn. This may merely be an inconvenience, but it may also force the user to grip the tool more tightly, once again probably increasing the risk of HAVS.

As a result of the above factors, antivibration gloves are not normally recommended as part of a programme of measures to reduce hand–arm vibration risks. They may keep the hands warm and this can help to protect against vibration damage. However, any glove will do this, and it would be more beneficial to choose a type that allows good control of the tool without an excessively tight grip. In any case, the antivibration types are considerably more expensive than most other gloves.

Case study 18.1 Managing hand–arm vibration exposure

An employee at Bradup Foundry has the job of fettling components and packing them on pallets ready for delivery to a particular customer. The job involves the use of a number of different grinding tools to cut off flash, grind specified surfaces to the required finish, and clean out holes. Because of the customer's quality requirements the job has always been carried out by one person who is familiar with exactly what is required. Very few components have been rejected in recent years, but a hand–arm vibration exposure survey has revealed that this employee is exposed to an equivalent continuous acceleration – A(8) – of between 9 and 10 ms^{-2}. It is now an urgent priority to reduce this exposure. Some hand–arm vibration measurements were carried out to test the effect of substituting different grinders which were thought likely to reduce vibration exposure. At the same time, the casting process was investigated to see what could be done to improve the quality of the castings. The customer was also contacted to discuss their precise requirements for incoming components. Improvements in the casting process and a change in one of the grinders used was found to reduce the value of A(8) for the original job to below 7 ms^{-2}. The customer, though, was reluctant to accept a reduction in component quality, and in fact was planning to raise the specification in the near future. The value of 7 ms^{-2} is still above the exposure limit value of the Physical Agents (Vibration) Directive and could be expected to result in symptoms of HAVS in a significant proportion of those exposed after a few years' exposure.

As an interim measure, the fettler had been working with an assistant so that the hand–arm vibration exposure of each was reduced. It was now decided that he would be given greater responsibility for the production and quality control of these components and would train the assistant to carry out the full range of fettling operations under supervision.

Both workers would in future be required to record the time spent using each type of grinder on a daily record sheet. These were collected by the health and safety manager who monitored the daily exposure of each using typical vibration magnitudes measured for each grinder. However, both workers were also warned of daily time limits for the use of each tool. Although these limits on individual tools did not take account of all the possible combinations of vibration exposure that could take place in a single day, it was thought that only a very simple system would be practicable to operate at shop-floor level.

Managers, meanwhile, entered into discussions with the customer about a long-term contract to supply several of the components to an improved specification. This would make it possible to invest in a more sophisticated casting process which would reduce the quantity of fettling required. This would in turn mean that the hand–arm vibration of the fettlers could be reduced to a more satisfactory long-term level, the target being to bring it below the exposure action value.

19

Controlling whole body vibration exposure

Approaches to the control of whole body vibration exposure

Chapter 15 discusses some of the management measures which can be used to work towards a reduction in the exposure of employees to any physical agent, whole body vibration included. The elimination of unnecessary tasks, the sharing of work involving exposure to hazards, and the use of available information to choose machinery involving lower levels of exposure are all relevant here. Some of the features of whole body vibration, and of the kinds of work in which it is a serious hazard, make it more difficult to apply these principles. For example, there may be a limited choice of vehicles to do a particular task, vehicles tend to have a longer life than small power tools, and replacement is very expensive. The high levels of skill involved may make it difficult to share the exposure between two or more drivers.

Although exposure can be reduced by choosing engines, or preparing terrain, in such a way that vibration is minimized at source, there are many vehicles for which the main way in which whole body vibration exposure can be controlled in most situations is by providing an appropriate seat. The vehicle manufacturer normally chooses the seat to be fitted, but in some cases a number of options will be available. If manufacturers are made aware of customer requirements, they will eventually choose to make better seat options available to suit the alternative uses to which their machines may be put.

This chapter first sets the choice of a seat to reduce whole body vibration exposure in an ergonomic context. Later, the ways in which the vibration reducing performance of a seat can be measured are described. The final part of the chapter looks at how shocks can be transmitted to the spine by inappropriate or wrongly adjusted seats.

Ergonomics and whole body vibration

The selection of an appropriate seat requires consideration of a number of factors to do with appropriate posture, visibility, the ability to operate controls and the physical environment. The consideration of these factors together constitutes an ergonomic assessment of a particular task or workstation. It is rare in buildings that vibration levels need to be considered explicitly as part of an ergonomic assessment, but in vehicles vibration is a key factor which interacts with many of the other assessment components.

Ergonomic assessments may be generic or they may be specific to an individual. A generic assessment aims to determine the suitability of a workstation for the normal range of human beings who can be expected to be deployed to that position, whereas a specific assessment will look at the individual and how well the physical arrangements are matched to that particular individual. During a generic assessment, it is clearly very important to check that various adjustments can easily be made to suit the normal range of human shapes, sizes and capabilities. For an individual assessment it would be necessary to check that appropriate adjustments had in fact been made and that appropriate additional equipment had been provided or modifications made.

For example, private cars have a driving seat which can be moved backwards and forwards as a whole, while the angle of the backrest can be adjusted separately. The height of the seat is often adjustable too, as is the height of the steering wheel. The mirrors can be angled in the vertical and horizontal planes. A new driver will normally make these adjustments before starting the engine, and it would be highly dangerous to travel in a car if this had not been done. The driver might have limited ability to see in front and behind, as well as a reduced ability to operate the controls correctly.

For good commercial reasons, car manufacturers aim to make their vehicles suitable for at least 90 per cent of the population, but even so not every car can be suitable for every possible driver. Commercial pressures mean that a great deal of effort goes into reducing vibration levels and to ensuring maximum comfort for the car's occupants. In the vehicle manufacturing industry these matters are referred to collectively as NVH (standing for noise, vibration and harshness) since actions intended to reduce noise and vibration and to ensure a smooth ride tend to interact with each other.

The pressures on a commercial vehicle designer are rather different from those operating in the domestic sector. For one thing the choice of vehicle will not be made by the person who ends up driving it, so more weight may be attached to economy and efficiency than to driver comfort. The range of tasks to be carried out by a single vehicle may be much greater. A tractor, for example, can be fitted with a number of different attachments working in different ways. A forklift truck must be able to manoeuvre in much tighter spaces than a private car and steering arrangements will need to be different as a result. As far as vibration is concerned, many commercial vehicles operate over much rougher terrain than private cars, and vibration reduction has in the past been a much lower priority than in cars.

Driven by greater awareness of health risks, and by legislation, the design and selection of seats on commercial vehicles is now receiving much greater attention than has been the case in the past. Vibration isolation is just one factor involved, alongside posture, visibility and the ability to manipulate controls. At very high vibration levels, vision and fine motor control are affected, and it must be remembered that the misinterpretation of a display or a loss of precision when manipulating tools could lead to serious safety consequences.

At levels much lower than those which interfere with vision and motor control, there is a possibility that whole body vibration exposure is a significant cause of back injuries. While the importance of vibration as a causative factor is controversial, it has been established that back problems are prevalent among those who spend a high proportion of their time operating off-road vehicles. If vibration is not the primary cause, then it may operate synergistically with other causes, among which poor posture is probably the most important.

Some sort of assessment of vibration exposure will therefore form part of any ergonomic assessment of the vehicle driving task. Off-road vehicles are particularly likely to impose undesirable vibration levels, but road vehicles may do so as well. Significant periods spent travelling by air – particularly in helicopters – or by train in the course of employment may also require an assessment. An initial assessment can be made using information published in databases or provided by the vehicle manufacturer, and in some cases this initial assessment will indicate that on-site measurements are required. In-house modifications to seats and suspensions are neither practical nor desirable. The outcome of an ergonomic assessment is likely to be one or more of the following:

- Limits on the time to be spent operating vehicles which expose occupants to high vibration levels;
- Selection of alternative optional seating arrangements available from the vehicle manufacturer;
- Specification to possible suppliers of the vibration performance required when purchasing new plant;
- Pressure on vehicle manufacturers to develop new vehicles, or modify existing models, to expose their occupants to lower levels of whole body vibration.

The types of seat normally fitted to commercial vehicles have developed over the years (Figure 19.1). In the mid-twentieth century most tractors had 'bucket' seats. These were rigid steel constructions which did nothing to isolate the driver from vibration and shock. On the other hand, there was no danger that a badly designed seat would increase potential vibration exposure. Over the years, it became common for a foam pad to be fitted to the seat; these are normally

Figure 19.1 (a) A basic seat on a small vehicle; (b) the underside of the same seat, showing the absence of any suspension. The spring in the centre is part of a safety interlock arrangement and does not support the seat.

effective at isolating higher frequencies, but are of very limited benefit at the frequencies of interest when considering the effects of whole body vibration.

To provide effective isolation against these frequencies more sophisticated, spring-mounted seats are required and these are now common on the more expensive vehicles (Figure 19.2). As long as the seat is carefully designed, taking into consideration the principles of vibration isolation, it may be able to reduce vibration levels by up to 50 per cent. The natural frequency of a standard sprung seat is difficult to reduce below about 4 Hz. This kind of seat will reduce the magnitude of vibrations having a frequency above 6 Hz, but those components below 6 Hz – which are significant in off-road vehicles – may actually be increased. To reduce low frequency vibration it is necessary to provide additional isolation effective at low frequencies. In large vehicles this can be done by introducing a second stage of isolation. Either the seat is mounted on a platform suspended in such a way that the natural frequency of this secondary suspension is much lower, or alternatively the cab itself benefits from a suspension system. The much greater mass involved then helps to achieve a lower natural frequency. The design of vehicle seats and cabs to reduce whole body vibration is discussed in detail by Lines and Stayner (2000).

The design changes described above are becoming standard on the larger, more expensive vehicles intended for use both on- and off-road. However, vehicles in use in industry, agriculture or construction may be many years old, or they may be cheaper models. At the lighter end of the market, there are a great many

Figure 19.2 A more sophisticated seat. Although the suspension mechanism is covered for safety reasons, the various adjustments available can be seen.

vehicles which are physically too small to be fitted with sophisticated seat arrangements. Vehicles such as street sweepers, ride-on lawn mowers and dump trucks are specifically designed to move in restricted spaces and with a high degree of manoeuvrability. Size considerations apart, the cost of developing low-vibration seats for this type of vehicle means that progress in reducing driver vibration levels is much slower.

Measuring the vibration isolation afforded by a seat

The isolation performance of a vehicle seat can be assessed by measuring the seat effective amplitude transmissibility or SEAT (pronounced see-at). SEAT measurements involve simultaneous vibration measurements on the vehicle floor and on the seat. The floor accelerometer must be rigidly mounted at the point of attachment of the seat, while the seat accelerometer will normally be part of a triaxial seat accelerometer set. In principle, a SEAT value can be measured along any axis, but in practice this is normally done for the z-axis.

It is possible to measure the transmissibility of the seat across the frequency range of interest. This is done by simply dividing the magnitude of the acceleration at the seat pad by the acceleration magnitude at the base of the seat. In so far as the seat and occupant can be modelled as a simple mass-spring system as shown in Figure 16.3, the transmissibility will vary approximately in a the same manner as is shown in Figure 16.4. A real seat will behave in a rather more complicated way than this, but the main features should still be observable:

- A transmissibility of 1 at very low frequencies;
- A high transmissibility around the natural frequency of the system;
- A transmissibility less than 1 at frequencies significantly greater than 1.4 times the natural frequency.

A graph or table of transmissibility versus frequency does not allow a quick judgement to be made of the quality of the seat, and neither does it indicate which frequency ranges are of most interest when assessing seat performance (since these also depend on the frequency components in the vibration transmitted from the ground). On the other hand, these kinds of data are properties of the seat itself, and will be the same – within the normal limits of measurement accuracy – whenever it is measured. They can indicate the particularly frequencies at which problems may occur.

The SEAT value (Figure 19.3) can be calculated from the transmissibility at various frequencies as long as the frequency weighting to be used is known. Alternatively it can be calculated directly from frequency weighted measurements. It is necessary for the method of time-averaging – either rms or VDV – to be decided since this also has an effect on the measured SEAT value. Essentially, the SEAT value is the ratio of the magnitude of the frequency weighted,

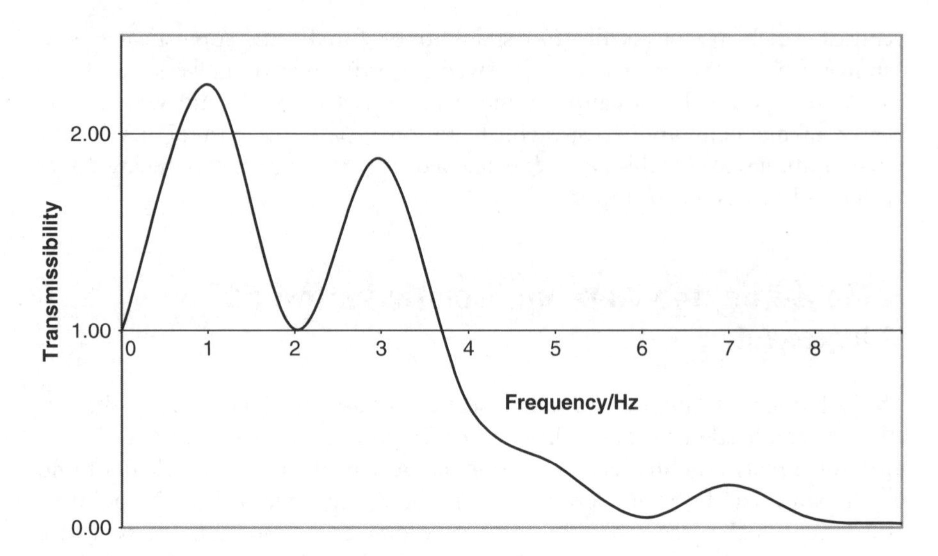

Figure 19.3 Transmissibility of a car seat (the measured SEAT value for this seat was 104 per cent).

time-averaged value at the seat pad to the magnitude using the same frequency weighting and time-averaging system at the point of attachment of the seat. The weighted values involved will depend on the spectrum of the input vibration, so that the measured SEAT value will also depend on the terrain over which the vehicle is driven, and the manner in which it is driven. Comparisons between seats and/or vehicles can only be carried out if a special test track is used.

$$SEAT_{rms} = \frac{\text{Frequency weighted rms-averaged acceleration on the seat}}{\text{Frequency weighted rms-averaged acceleration at the base of the seat}} \times 100\% \tag{19.1}$$

$$SEAT_{VDV} = \frac{\text{Frequency weighted VDV on the seat}}{\text{Frequency weighted VDV at the base of the seat}} \times 100\% \tag{19.2}$$

A SEAT value of less than 100 per cent implies that the vibration exposure is less than would be the case if a completely rigid seat was in use. Values greater than 100 per cent imply that the vibration exposure is worse than if a completely rigid seat was in use.

One study (Paddan and Griffin, 2001) measured SEAT values on 100 vehicles and found values ranging from 32 to 160 per cent on 100 vehicles of different types, and a SEAT value of 250 per cent on a helicopter seat. Moreover, the authors then carried out a theoretical analysis of the effects of swapping seats between different vehicles used in their study and showed that in 94 cases out of 100 this would result in an improvement of the seat's performance. This indicates in general terms the potential for improving the vibration isolation of vehicle seats without waiting for major advances in seat technology. It does not, of course, mean that it would be practical to exchange the actual seats concerned in most cases.

Seat over-travel and shock

The degree of movement that is available in any suspension system is limited by the physical characteristics of its components. There is a limit to the distance that any seat suspension can travel either up or down from its equilibrium position. The seat designer will normally arrange for the movement of the seat to be constrained within certain limits beyond which springs would be overstressed or linkages would break. There is in any case a need to limit the movement of the seat in conditions where the vibration applied is close to the natural frequency of the system comprising the seat and its occupant.

The simplest way of limiting seat movement is by putting a physical end-stop at each end of the permitted travel. Under normal circumstances it is intended that the seat will not strike this end stop, but if for any reason it does, a shock will be transmitted to the body of the occupant. This type of shock, although very short lived, can contain high-amplitude, high-frequency components which are potentially responsible for a much greater degree of damage to the spine than exposure to continuous vibration at a much lower amplitude.

Seat over-travel is the name given to the operation of the seat outside its intended range. If it is moving outside this range it may or may not go so far as to impact the end-stops. There are two reasons why this may happen:

1. The seat is being exposed to vibration amplitudes which are higher than those it has been designed to protect against. Either the seat was an inappropriate choice on the part of the machine manufacturer, or the machine itself is being operated in conditions for which it was not intended; for example, a fork-lift truck designed for use on hard surfaces which is being operated over rough ground.
2. The seat has not been adjusted to suit its occupant. The applied vibration is likely to expose the seat equally to motion both upwards and downwards from its equilibrium position. To cope with the greatest vibration amplitude, it should start off in a central position so that it can cope equally with motion in both directions. The static position will depend on the weight of the operator, so that a sprung seat should incorporate a means by which the operator can adjust the seat to take account of his or her weight. If this adjustment is not made, then the occupant may be exposed to damaging shocks.

Any whole body vibration figures declared by the manufacturer will assume correct adjustment of the seat. In the event that it is not correctly adjusted, the actual vibration magnitude may be several times greater.

Seat designers have developed various ways of reducing the likelihood of end-stop impacts. They range from the use of resilient buffers on the end-stops to sophisticated springing systems which can absorb much greater magnitudes of applied vibration. None of these approaches are suitable for in-house modification of existing vehicles. The key strategies to guard against high levels of WBV exposure due to end-stop impacts are:

- Use of an appropriate vehicle for the task;
- Training of operators in the correct way to adjust the seat;
- Management procedures to include checks that the adjustments have in fact been made;
- Encouraging operators to report if end-stop impacts are taking place;
- Making vehicle suppliers aware of the conditions under which their products are to be used, and of the acceptable level of vibration exposure.

End-stop impact has been discussed above in the context of sprung vehicle seats. Similar effects can arise within the vehicle's suspension system if vehicles are used in conditions for which they were not designed, for example, in the case of a private car driven fast down a farm track.

Appendix A: Frequency weighting values

Frequency weightings for noise measurement

Frequency/Hz	A weighting/dB	C weighting/dB	Z weighting/dB	Tolerance for class 1 SLMs	Tolerance for class 2 SLMs
16	−56.7	−8.5	0	+2.5/−4.5 dB	+5.5/−∞ dB
31.5	−39.4	−3.0	0	±2.0 dB	±3.5 dB
63	−26.2	−0.8	0	±1.5 dB	±2.5 dB
125	−16.1	−0.2	0	±1.5 dB	±2.0 dB
250	−8.6	0	0	±1.4 dB	±1.9 dB
500	−3.2	0	0	±1.4 dB	±1.9 dB
1000	0	0	0	±1.1 dB	±1.4 dB
2000	1.2	−0.2	0	±1.6 dB	±2.6 dB
4000	1.0	−0.8	0	±1.6 dB	±3.6 dB
8000	−1.1	−3.0	0	+2.1/−3.1 dB	±5.6 dB
16 000	−6.6	−8.5	0	+3.5/−17.0 dB	+6.0/−∞ dB

Frequency weightings for vibration measurement

Type of vibration	Hand–arm vibration		Whole body vibration		
Name	W_h		W_k	W_d	
Axes	All		Vertical	Horizontal	
Frequency/Hz	Weighting	Tolerance	Weighting	Weighting	Tolerance
1	0.025		0.87	0.43	+26%/−21%
1.25	0.040		0.95	0.51	
1.6	0.063	+26%/−∞	0.96	0.59	
2	0.10		0.89	0.68	
2.5	0.16		0.78	0.79	+12%/−11%
3.15	0.25		0.64	0.91	
4	0.38		0.52	1.01	
5	0.55		0.41	1.06	
6.3	0.73	+26%/−21%	0.32	1.02	
8	0.87		0.25	0.91	0
10	0.96		0.20	0.76	
12.5	0.96		0.16	0.62	
16	0.90		0.13	0.50	
20	0.78		0.10	0.40	
25	0.65		0.079	0.32	
31.5	0.52	+12%/−11%	0.063	0.25	+12%/−11%
40	0.41		0.049	0.20	
50	0.32		0.039	0.15	
63	0.26		0.029	0.12	
80	0.20	0	0.021	0.085	
100	0.16		0.014	0.056	+26%/−21%
125	0.13		0.008	0.04	
160	0.10		0.005	0.019	
200	0.080	+12%/−11%	0.0024	0.01	
250	0.063		0.0012	0.005	+26%/−∞
315	0.050		0.00063	0.003	
400	0.039		0.0005	0.001	
500	0.031		0.0004	0.001	

Appendix B: Action levels and duties

The Noise at Work Regulations 1989

Levels

Name	Quantity involved	Level
First action level	$L_{AEP,d}$	85 dB
Second action level	$L_{AEP,d}$	90 dB
Peak action level	L_{peak} 'unweighted'	140 dB or 200 Pa

General Employer duties

- Reduce the risk of hearing damage to the lowest level reasonably practicable.

Employer duties when it is likely that either the first or the peak action level is exceeded

- Arrange for a competent person to carry out a noise exposure assessment
- Keep a record of any such assessment until a further assessment is carried out.

Employer duties when the first action level is exceeded

- Provide employees with hearing protection capable of reducing noise exposure below the second and/or peak action level
- Provide hearing protection if requested

- Ensure that equipment (other than hearing protection) provided to reduce noise exposure is used
- Ensure that noise reduction equipment (including hearing protection) is maintained
- Provide training and information on the risk of hearing damage, ways of minimizing that risk, how to obtain hearing protection, and the employee's own duties.

Employer duties when the second or peak action level is exceeded

- Reduce exposure of employees to noise, so far as is reasonably practicable, by means other than hearing protection
- Ensure that hearing protection is used
- Define and mark ear protection zones, and ensure that any employee entering such a zone is wearing hearing protection.

Employee duties

- Wear hearing protection if exposure is above the second or peak action levels
- Use other noise reduction equipment when provided
- Report any defects in hearing protection and other noise reduction equipment.

Other duties

- Suppliers of work equipment have a duty to provide information about the noise likely to be generated by it.

Implementation

- Directive 86/188/EEC was issued on 12 May 1986.
- The Noise at Work Regulations 1989 came into force in the UK on 1 January 1990.

The Physical Agents (Noise) Directive

Levels

Name	Quantity involved	Level	Hearing protection
Lower exposure action value	$L_{EX,8\ \mathrm{hours}}$ (equivalent to $L_{AEP,d}$)	80 dB	Not taken into account when assessing the dose
Upper exposure action value		85 dB	
Exposure limit value		87 dB	Taken into account
Lower peak exposure action value	L_{Cpeak}	135 dB or 112 Pa	Not taken into account when assessing the dose
Upper peak exposure action value		137 dB or 140 Pa	
Peak exposure limit value		140 dB or 200 Pa	Taken into account

General employer duties

- Assess, and if necessary measure, the levels of noise to which workers are exposed
- Consult with employees
- Eliminate or minimize any risks to health due to noise exposure.

Employer duties when exposure is greater than the lower exposure action value (EAV)

- Make personal hearing protection available
- Provide information and training to employees about the risks resulting from exposure to noise, how to recognize hearing damage, the results of noise exposure assessments, the correct use of hearing protection and safe working practices to minimize noise exposure.

Employer duties when exposure is greater than the upper exposure action value (EAV)

- Ensure that personal hearing protection is used
- Establish and mark hearing protection zones
- Arrange hearing tests.

Employer duties when exposure is greater than the exposure limit value (ELV)

- Take action to reduce exposure below the ELV
- Take action to prevent a recurrence of exposure above the ELV.

Implementation

- Published in the Official Journal of the European Union on 15 February 2003
- Member states are required to implement the directive into domestic legislation by 15 February 2006
- For most employees, the directive becomes effective as soon as the domestic legislation comes into force
- For the music and entertainment sectors, implementation may be delayed until February 2008
- For personnel on seagoing ships, implementation may be delayed until February 2011.

The Physical Agents (Vibration) Directive

Levels for hand–arm vibration

Name	Quantity involved	Level
Exposure action value	A(8), also called $a_{hv(eq,8h)}$	2.5 ms^{-2}
Exposure limit value		5.0 ms^{-2}

Levels for whole body vibration

Name	Quantity involved	Level
Exposure action value	Equivalent 8-h continuous level	0.5 ms^{-2}
Exposure limit value	Equivalent 8-h continuous level	1.15 ms^{-2}

or

Quantity involved	Level
VDV	9.1 ms$^{-1.75}$
VDV	21 ms$^{-1.75}$

Member states must decide whether to frame their domestic legislation on whole body vibration in terms of A(8) or VDV exposure action and limit values.

General employer duties

- Assess, and if necessary measure, the levels of vibration to which workers are exposed
- Consultation with employees
- Eliminate or minimize any risks to health due to vibration exposure.

Employer duties when exposure is greater than the exposure action value

- Reduce to a minimum employee exposure to vibration and the attendant risks, by technical and/or organizational means
- Information and training of employees, to cover the exposure action and limit values, the outcome of the exposure assessment and the potential injuries involved, ways to detect and report signs of injury, and safe working practices to minimize vibration exposure
- Health surveillance.

Employer duties when exposure is greater than the exposure limit value

- Take action to reduce exposure below the ELV
- Take action to prevent a recurrence of exposure above the ELV.

Implementation

- Published in the Official Journal of the European Union on 6 July 2002
- Member states are required to implement the directive into domestic legislation by 6 July 2005
- For most employees, the directive becomes effective as soon as the domestic legislation comes into force
- Implementation may be delayed until 6 July 2010 as regards exposure to vibration from equipment brought into use before 6 July 2007
- For the agriculture and forestry sectors, implementation may be delayed until 6 July 2014.

Appendix C: Glossary of symbols and abbreviations

Abbreviation	Explanation	Comments
%age dose	Formerly used to assess daily noise exposure	Originally an exposure of 100% was equivalent to an $L_{EP,d}$ of 90 dB. May cause confusion due to the different action levels now in use.
A(4)	The equivalent 4-h continuous acceleration	This is the acceleration to which an employee would need to be exposed continuously for 4 h in order to receive the same amount of vibrational energy as is received from the actual, fluctuating vibration exposure. This was used in the 1989 version of ISO 5349.
A(8)	The equivalent 8-h continuous acceleration	Used for assessing daily hand–arm vibration dose, this is the acceleration to which an employee would need to be exposed continuously for 8 h in order to receive the same amount of vibrational energy as is received from the actual, fluctuating, vibration exposure.
$a_{h,w}$	In BS 6842:1987 (now withdrawn), this symbol was used for the combined axis instantaneous acceleration measured at a tool handle	In ISO 5349:2001, it is replaced by a_{hv}.
a_{hv}	The hand–arm weighted acceleration at a tool handle, summed over the three measurement axes	Defined in ISO 5349:2001.
$a_{hv(eq,8h)}$	Introduced in ISO 5349:2001 as an alternative notation for A(8).	
a_{hw}	A former abbreviation for the hand–arm weighted, combined axis, acceleration measured at a tool handle. Defined in BS 6842:1987	Now replaced in ISO 5349:2001 by a_{hv}.
a_{hwx} a_{hwy} a_{hwz}	The hand–arm weighted acceleration measured at a tool handle along the x (or y, or z) axis.	Defined in ISO 5349:2001.

Abbreviation	Explanation	Comments
$a_{x,h,w}$ $a_{y,h,w}$ $a_{z,h,w}$	In the former BS 6842:1987, these were the symbols used for the x, y and z components of the hand–arm weighted acceleration levels measured at a tool handle	They are replaced in ISO 5349:2001 by a_{hx}, a_{hy}, and a_{hz}.
APL	Assumed protected level	In predictions of the effectiveness of hearing protectors, the sound pressure levels predicted at the ear.
BSI	British Standards Institution	The body which publishes standards for use in the UK.
dB	The decibel	
dB(A)	Decibels, measured using A weighting	An obsolete notation, still frequently used.
dB(C)	Decibels, measured using C weighting	As above.
E_A	The sound exposure	Measured in pascals squared-hours.
EAV	Exposure action value	Defined in the Physical Agents Directives.
ELV	Exposure limit value	Defined in the Physical Agents Directives.
EPZ	Ear protection zone	See HPZ.
EU	European Union	Formerly EC (European Community), EEC (European Economic Community).
eVDV	Estimated vibration dose value	A quantity calculated as an approximation to the VDV if only rms averaging measurement equipment is available.
g	The acceleration due to gravity at the Earth's surface	Approximately equal to $9.8\,ms^{-2}$;$10\,ms^{-2}$ is sometimes used as an approximation.
H	High frequency attenuation	Used in the HML method of calculating hearing protector effectiveness; defined in ISO 28469 Part 2:1992.
HASAWA	The Health and Safety at Work, etc., Act, 1974	The primary legislation under which the Noise at Work and Vibration at Work Regulations are issued.
HATS	Head and torso simulator	Dummy head used for some measurements of noise exposure from headsets.
HAV	Hand–arm vibration	
HAVS	Hand–arm vibration syndrome	The set of vascular, sensorineural and musculoskeletal symptoms arising from long-term exposure of the hand/arm to vibration.
HML	High-medium-low	A method for predicting the effectiveness of hearing protection.
HP	Hearing protection	
HPZ	Hearing protection zone	Part of a workplace where hearing protection is compulsory even for those spending a short period there, see EPZ.
HSE	The Health and Safety Executive	The agency responsible for the enforcement of health and safety legislation in the UK.
Hz	The hertz	The SI unit of frequency; formerly the cycle per second.
IEC	International Electrotechnical Commission	This Geneva-based body develops standards for certain types of measuring equipment.
ISO	International Standards Organization	This Geneva-based body develops standards which may or may not be adopted by national standards organizations such as BSI.
L	Low frequency attenuation	Used in the HML method of calculating hearing protector effectiveness; defined in ISO 28469 Part 2:1992.
L_A	Sound pressure level, measured using A weighting	If the level is steady, this is equivalent to L_{Aeq}.

Continued

Abbreviation	Explanation	Comments
L_{AE}	The sound exposure level	Used in predicting long-term exposure due to a number of short, noisy events. Also known as single event level, L_{eA}, L_{ax}, SEL.
$L_{AEP,d}$	Personal daily noise exposure	The sound pressure level, averaged and normalized over a standard 8-h shift. The quantity to which the first and second action levels of the NAWR relate.
L_{Aeq}	Equivalent continuous level, measured using A weighting	
$L_{Aeq,t}$	Equivalent continuous level, measured using A weighting and over a specified time	
L_{Afp}	Sound pressure level measured using fast time constant and A weighting	
L_{ax}	See L_{AE}	
L_C	Sound pressure level, measured using C weighting	
L_{Cpk}	The peak, C weighted, sound pressure level	Used for the peak action level of the NAWR.
L_{eA}	See L_{AE}	
$L_{EP,w}$	The weekly personal noise exposure	The daily $L_{EP,d}$ averaged over a full working week.
$L_{EP,d}$	Personal daily noise exposure	The same as $L_{AEP,d}$.
L_{eq}	The equivalent continuous noise level	Frequency weighting needs to be specified.
$L_{EX,8\text{ hours}}$	The equivalent 8-h level	Equivalent to $L_{EP,d}$. This terminology is used in the Physical Agents (Noise) Directive.
L_I	Sound intensity level	A decibel quantity based on sound intensity.
L_{max}	The maximum rms sound pressure level during a specified measurement period	The value of L_{max} depends on whether fast or slow time constant is selected. Not used in workplace noise assessment.
L_p	Sound pressure level	Sometimes referred to as SPL.
L_{pA}	Sound pressure level, A weighted	The same as L_A.
L_{peak}	Peak sound pressure level	Normally the highest peak sound pressure level occurring during a specified measurement period. The frequency weighting also needs to be specified. Sometimes L_{peak} is used to denote the highest peak level occurring in rolling 1-s periods.
L_{pkmax}	Maximum peak sound pressure level	When L_{peak} is used to denote the highest peak level occurring in rolling 1-s periods, L_{pkmax} is the highest peak level during the entire measurement period.
L_W	Sound power level	A decibel quantity based on the sound power A weighted L_W values must be measured and declared for many noisy machines.
L_Z	The sound pressure level, measured using Z weighting	
M	Medium frequency attenuation	Used in the HML method of calculating hearing protector effectiveness; defined in ISO 28469 Part 2.
m	The metre	The unit of length in the SI system.
ms^{-1}	The metre per second	The SI unit of velocity.
$ms^{-1.75}$	Metres-seconds to the minus 1.75	The unit in which VDV is measured.
ms^{-2}	The metre per second squared	The SI unit of acceleration.
MTVV	Maximum transient vibration value	Defined in ISO 2631 Part 1:1997; used in one method for assessing whole body vibration.
NAWR	The Noise at Work Regulations 1989	Regulations which implement in the UK the European Directive 86/188.
NIPTS	Noise induced permanent threshold shift	A permanent hearing loss caused by noise exposure.

Abbreviation	Explanation	Comments
NRR	Noise reduction rating	A single-figure quantity used in the United States for hearing protector calculations. Similar to, but not the same as, SNR.
Pa	The pascal	The SI unit of pressure, equal to a force of one newton acting on each square metre of a surface.
PA(N)D	The Physical Agents (Noise) Directive	
PA(V)D	The Physical Agents (Vibration) Directive	
Pa^2-h	Pascals squared-hours	The unit in which the noise exposure, E_A, is measured
PNR	Predicted noise reduction	The calculated attenuation of a hearing protector using the HML method.
PPE	Personal protective equipment	
PWL	See L_W	
rmq	Root-mean-quad	The time averaging procedure used when measuring VDV.
rms	Root-mean-square	A method of averaging alternating signals over time in a way which preserves the energy content of the original signal.
rss	Root-sum-of-squares	A method of adding different contributions (either different frequency components or contributions from different measurement axes) of vibration so as to arrive at an overall energy sum.
SEAT	Seat effective amplitude transmissibility	A measure of the performance of a seat in isolating its occupant from vibration.
SEL	See L_{AE}	
SI	Système Internationale	An agreed international system of units for most types of measurement, based on the use of the metre, the kilogram and the second to measure length, mass and time, respectively.
SLM	Sound level meter	
SNR	Single number rating	A figure supplied by a hearing protection supplier, and used in the simplest method for predicting the protected level.
SPL	Sound pressure level	The preferred terminology is now L_p.
SWL	See L_W	
TTS	Temporary threshold shift	A temporary hearing loss caused by noise exposure, but which is reversed over the 2–3 days following noise exposure.
VDV	Vibration dose value	One quantity used to assess the dose of whole-body vibration. Uses rmq averaging. Measured in units of $ms^{-1.75}$.
W	Sound power	The power (energy per second) radiated as noise by a noise source. A characteristic of that source. Measured in watts.
W	Watt	The SI unit of power (including sound power).
W_b		Frequency weighting defined in BS 6841:1987, and used to assess vibration exposure along the vertical axis (replaces W_k in this standard).
W_c		Frequency weighting defined in ISO 2631:1997 and also in BS 6841:1987, and used to assess vibration transmitted via a backrest.
W_d	The most important frequency weighting used to assess vibration exposure along the two horizontal axes	Defined in ISO 2631:1997. Also in BS 6841:1987.

Continued

Abbreviation	Explanation	Comments
W_e		Frequency weighting defined in ISO 2631:1997 and also in BS 6841:1987, and used to assess rotational vibration exposure.
W_f		Frequency weighting defined in ISO 2631:1997 and also in BS 6841:1987, and used to assess the likelihood of motion sickness.
W_g		Frequency weighting defined in BS 6841:1987, and used to assess some effects of vertical axis vibration.
W_h	The hand–arm vibration frequency weighting	Defined in ISO 5349 Part 1:2001.
W_k	The most important frequency weighting used to assess some effects of vertical axis vibration	Defined in ISO 2631 Part 1:1997.
W_j		Frequency weighting defined in ISO 2631 Part 1:1997, and used to assess some effects of vertical axis vibration on a recumbent subject.
WBV	Whole body vibration	
Wm^{-2}	Watts per metre squared	The SI unit of sound intensity.
x,y,z	x, y and z axes	Traditionally, a set of three co-ordinate axes are called the x, y and z axes, respectively. The relevant standards specify their orientation for hand–arm and whole body vibration.

Appendix D: Further calculation examples and answers

Questions

Noise

1. An employee is exposed to the following noise levels for different periods of the working day. (a) Work out the L_{eq} and the $L_{EP,d}$ for the whole shift, and (b) decide the order in which the different periods contribute to the daily noise exposure.

Period	SPL/dB	Time
A	98	30 min
B	93	6 h
C	87	2½ h

2. An employee is exposed to the following noise levels for different periods of the working day. (a) Work out the L_{eq} and the $L_{EP,d}$ for the whole shift, and (b) decide the order in which the different periods contribute to the daily noise exposure.

Period	SPL/dB	Time
A	108	10 min
B	88	2 h
C	86	1½ h
D	94	1 h
E	90	30 min

3. Measurements made during blasting operations at a quarry show that a particular employee is exposed to an L_{AE} of 121 dB. (a) What is this person's daily exposure if only one such operation takes place during the day? (b) What is the same employee's $L_{EP,d}$ if a military aircraft also passes overhead, producing an L_{AE} of 119 dB?
4. The A weighted SPL at a particular workstation is measured at 92 dB. How long can an unprotected employee be exposed to this level before the $L_{EP,d}$ reaches the following levels: (a) 80 dB; (b) 85 dB; (c) 87 dB; (d) 90 dB?
5. An employee is exposed to an SPL of 97 dB for a period of 30 min. For how long during the same working day can this employee be exposed to a level of 94 dB before $L_{EP,d}$ reaches 87 dB?
6. A second employee has also been exposed to an SPL of 97 dB for 30 min. What is the maximum SPL to which this employee can be exposed for the remainder of the 8-h shift before his or her $L_{EP,d}$ exceeds 87 dB?
7. A manufacturer provides the following attenuation data with a particular model of hearing protector.

Band/Hz	63	125	250	500	1 k	2 k	4 k	8 k
Mean attenuation/dB	14.0	13.1	11.9	18.6	29.0	29.7	37.1	31.9
Standard deviation/dB	5.0	3.6	2.4	2.7	3.0	3.2	4.0	6.3
	H	28	M	19	L	13	SNR	23

It is proposed to use these hearing protectors in an environment for which the following SPL measurements have been made. Predict the protected SPL using (a) the octave band method; (b) the HML method; and (c) the SNR method.

Band/Hz	63	125	250	500	1 k	2 k	4 k	8 k	A	C
Band SPL/dB	84.3	91.2	93.7	89.1	86.0	83.0	78.6	64.9	91.8	97.2

8. It is proposed to use the same hearing protectors in the following noise environment. Calculate the protected levels to be expected using the three methods. What time limit should be imposed on workers using this protection if it is required to keep their daily exposure below 87 dB?

Band/Hz	63	125	250	500	1 k	2 k	4 k	8 k	A	C
SPL/dB	81.2	88.6	101.1	109.2	109.7	108.4	104.0	91.4	113.6	114.4

9. An employee is exposed during an 8-h shift to an unprotected noise level of 106 dB, and has been provided with hearing protection which reduces this to 79 dB. What will his daily dose, $L_{EP,d}$, be if he wears this hearing protection for (a) the whole of the shift; (b) all but 10 min of the shift; and (c) 7 of the 8 h?

Hand–arm vibration

10. The following hand–arm weighted accelerations are measured at a tool handle:

$$a_{hwx} = 2.9\,\text{ms}^{-2}$$
$$a_{hwy} = 4.3\,\text{ms}^{-2}$$
$$a_{hwz} = 3.5\,\text{ms}^{-2}$$

What is the combined axis acceleration?

11. A combined axes hand–arm acceleration of 6.0 ms^{-2} is measured at a chain saw handle. What is the operator's daily equivalent exposure if the tool is used for 2.5 h per day?
12. An employee spends the following times using vibrating tools in the course of a shift. Calculate the employee's 8-h equivalent exposure, A(8).

Operation	Combined axes acceleration	Time
Chipping	8.4 ms^{-2}	30 min
Grinding	5.2 ms^{-2}	2 h
Polishing	3.8 ms^{-2}	2 h

13. An employee spends the following times using vibrating tools in the course of a shift. (a) Calculate the employee's 8-h equivalent exposure, A(8). (b) Put the various operations in order of their contribution to the daily dose.

Operation	Combined axes acceleration	Time
Strimming	5.4 ms^{-2}	2 h
Leaf blowing	4.1 ms^{-2}	2 h
Hedge trimming	7.5 ms^{-2}	30 min

14. A tool exposes its operator to a combined axes hand–arm weighted acceleration of 7.5 ms^{-2}. What length of exposure would be required before (a) the exposure action value of 2.5 ms^{-2}; and (b) the exposure limit value of 5.0 ms^{-2} were reached?
15. The following accelerations are measured for each axis on a tool handle. What daily time limit will ensure that the operator is not exposed above the exposure limit value?

x-axis	$3.2\,ms^{-2}$
y-axis	$3.4\,ms^{-2}$
z-axis	$5.0\,ms^{-2}$

16. An employee has used a chipping hammer which entailed exposure to an acceleration of $7.5\,ms^{-2}$ for 30 min. What is the maximum time for which this employee can now use a grinder at whose handle the acceleration is $4.3\,ms^{-2}$ before the exposure action value is reached?

Whole body vibration

17. A delivery driver is exposed to the following weighted rms acceleration levels during a typical 5-h run. Calculate the equivalent 8-h levels for this shift.

x-axis	$0.52\,ms^{-2}$
y-axis	$0.43\,ms^{-2}$
z-axis	$0.75\,ms^{-2}$

18. A digger operator is exposed to a VDV of $7.2\,ms^{-1.75}$ (x-axis), $10.5\,ms^{-1.75}$ (y-axis), and $11.3\,ms^{-1.75}$ (z-axis) during a cycle of operations which involves digging 10 m of trench and then carrying out other duties not involving WBV exposure. In a day, the driver will normally go through three such cycles. What would the daily VDVs be?
19. Over a typical journey, a bus driver is exposed to the following time-averaged, frequency weighted vibration levels. Work out the equivalent 8-h level and the eVDV for this driver, given that a normal day involves 7 h of driving.

x-axis	$0.41\,ms^{-2}$
y-axis	$0.24\,ms^{-2}$
z-axis	$0.60\,ms^{-2}$

Answers

1. $L_{eq} = 93.1$ dB; $L_{EP,d} = 92.8$ dB; B > A > C.
2. $L_{eq} = 94.4$ dB; $L_{EP,d} = 92.5$ dB; A > D > B > C > E.
3. (a) 76.4 dB; (b) 78.5 dB.
4.

$L_{EP,d}$	Time taken
80 dB	0.5 h
85 dB	1.6 h
87 dB	2.5 h
90 dB	5.0 h

5. 33 min.
6. 83 dB.
7. (a) 77.1; (b) 75.3; (c) 74.2.
8. Octave band: protected level = 92.3 dB, time limit = 2.4 h.
   HML: protected level = 91.9 dB, time limit = 2.6 h.
   SNR: protected level = 91.4 dB, time limit = 2.9 h.
9. (a) 79 dB; (b) 89.6 dB; (c) 97.0 dB.
10. 6.3 ms^{-2}.
11. 3.4 ms^{-2}.
12. 3.8 ms^{-2}.
13. (a) 3.9 ms^{-2}; (b) Strimming (partial A(8) 2.7 ms^{-2}), hedge trimming (partial A(8) 2.1 ms^{-2}), leaf blowing (partial A(8) 1.9 ms^{-2}).
14. (a) 54 min; (b) 3.6 h.
15. 4 h 16 min.
16. 1 h 11 min.
17. x-axis 0.58 ms^{-2}
    y-axis 0.48 ms^{-2}
    z-axis 0.59 ms^{-2}
18. x-axis 13.2 $ms^{-1.75}$, y-axis 19.3 $ms^{-1.75}$, z-axis 14.9 $ms^{-1.75}$.
19.

	A(8)	eVDV
x-axis	0.54 ms^{-2}	10.1
y-axis	0.31 ms^{-2}	5.9
z-axis	0.56 ms^{-2}	10.6

Bibliography

Barr, T. Enquiry into the effects of loud sounds upon the hearing of boilermakers and others who work amid noisy surroundings. *Proceedings of the Glasgow Philosophical Society* **17**, 223–230, 1886.

BS 5330:1976. Method of test for estimating the risk of hearing handicap due to noise exposure.

BS 6841:1987. Measurement and evaluation of human exposure to whole-body mechanical vibration and repeated shock.

BS 6842:1987. Measurement and evaluation of human exposure to vibration transmitted to the hand.

BS 7482 Part 1:1991. Instrumentation for the measurement of vibration exposure of human beings. Part 1: Specification for general requirements for instrumentation for measuring the vibration applied to human beings.

BS 7482 Part 2:1991. Instrumentation for the measurement of vibration exposure of human beings. Part 2: Specification for instrumentation for measuring vibration transmitted to the hand.

BS 7482 Part 3:1991. Instrumentation for the measurement of vibration exposure of human beings. Part 3: Specification for instrumentation for measuring vibration exposure to the whole body.

BS 7580 Part 1:1997. Specification for the verification of sound level meters. Comprehensive procedure.

BS 7580 Part 2:1997. Specification for the verification of sound level meters. Shortened procedure for type 2 sound level meters.

BS EN 352-1:2002 Hearing protectors. Part 1: Safety requirements and testing. Ear-muffs.

BS EN 352-2:2002 Hearing protectors. Part 2: Safety requirements and testing. Ear-plugs.

BS EN 352-3:2002 Hearing protectors. Part 3: Safety requirements and testing. Ear-muffs attached to an industrial safety helmet.

BS EN 352-4:2001 Hearing protectors. Part 4: Safety requirements and testing. Level-dependent ear-muffs.

BS EN 352-5:2002 Hearing protectors. Part 5: Safety requirements and testing. Active noise reduction ear-muffs.

BS EN 352-6:2002 Hearing protectors. Part 6: Safety requirements and testing. Ear-muffs with electrical audio input.

BS EN 352-7:2002 Hearing protectors. Part 7: Safety requirements and testing. Level-dependent ear-plugs.

BS EN 458:1994 Hearing protectors. Recommendations for selection, use, care and maintenance. Guidance document.

Carling, C. Hand–arm vibration syndrome: the legal aspects. In: Proud, G., Hodgson, C., Lees, T. (eds). *Hand–arm vibration syndrome*. Proceedings of the Joint Symposium. Leeds: H&H Scientific Consultants Ltd, 1999.

Committee on the Problem of Noise. Noise – final report. London: HMSO, 1963 (known as the Wilson Report).

Department of Employment. Code of practice for reducing the exposure of employed persons to noise. London: HMSO, 1972.

European Community. Council Directive 86/188/EEC on the protection of workers from the risks related to exposure to noise at work. 1986.

European Union. Directive 2002/44/EC of the European Parliament and of the Council of 25 June 2002 on the minimum health and safety requirements regarding the exposure of workers to the risks arising from physical agents (vibration).

European Union. Directive 2003/10/EC of the European Parliament and of the Council of 6 February 2003 on the minimum health and safety requirements regarding the exposure of workers to the risks arising from physical agents (noise).

Griffin, M.J. *Handbook of human vibration*. Academic Press, 1990.

Hamilton, A. A study of spastic anaemia in the hands of stonecutters. US Department of Labor, Bureau of Labor Statistics Bulletin **236**, 1918.

Health and Safety Commission. Proposals for new Control of Vibration at Work Regulations implementing the Physical Agents (Vibration) Directive (2002/44/EC). Hand–arm Vibration. Consultative Document CD190, 2003.

Health and Safety Commission. Proposals for new Control of Vibration at Work Regulations implementing the Physical Agents (Vibration) Directive (2002/44/EC). Whole Body Vibration. Consultative Document CD191, 2003.

Health and Safety Executive. HS(G)88; Hand–arm vibration. Sudbury: HSE Books, 1994.

Health and Safety Executive. Guidance note MS 26. A guide to audiometric testing programmes. Sudbury: HSE Books, 1995.

Health and Safety Executive. L108. Reducing noise at work. Sudbury: HSE Books, 1998.

Hewitt, S.M., Smeatham, D. Comparison of vibration emission data with vibration in use; final report. NV/00/11. Buxton: Health and Safety Laboratory, 2000.

IEC 60645 Part 1:2001. Electroacoustics – Audiological equipment. Part 1: Pure-tone audiometers (published as BS EN 606445-1:2001).

IEC 60651:1994. Specification for sound level meters (now withdrawn).

IEC 60804:2000. Integrating-averaging sound level meters (now withdrawn).

IEC 60942:2003. Electroacoustics. Sound calibrators (published as BS EN 60942:2003).

IEC 61672:1:2003. Electroacoustics. Sound level meters. Specifications.

ISO 226:2003. Acoustics – Normal equal loudness contours (published as BS ISO 226:2003).

ISO 389 Part 1:2000. Acoustics – Reference zero for the calibration of audiometric equipment. Part 1: Reference equivalent threshold sound pressure levels for pure tones and supra-aural earphones (published as BS EN ISO 389-1:2000).

ISO 2631 Part 1:1997. Mechanical vibration and shock – Evaluation of human exposure to whole body vibration. Part 1: General requirements, second edition.

ISO 2631 Part 2:2003. Mechanical vibration and shock – Evaluation of human exposure to whole-body vibration – Part 2: Vibration in buildings (1 Hz to 80 Hz).

ISO 4869-1:1993, ISO 4869-1:1990 Acoustics. Hearing protectors. Sound attenuation of hearing protectors. Subjective method of measurement.

ISO 4869-2:1995. Acoustics. Hearing protectors. Estimation of effective A-weighted sound pressure levels when hearing protectors are worn.

ISO 5349 Part 1:2001. Mechanical vibration – Measurement and assessment of human exposure to hand-transmitted vibration. Part 1: General guidelines (published as BS EN ISO 5349-1:2001).

ISO 5349 Part 2:2002. Mechanical vibration – Measurement and assessment of human exposure to hand-transmitted vibration. Part 2: Practical guidance for measurement at the workplace (published as BS EN ISO 5349-2:2002).

ISO 7505 Part 8:1986. Chain saws; Method of measurement of hand-transmitted vibration (published as BS 6916 Part 8:1988).

ISO 8041:1990. Human response to vibration. Measuring instrumentation (also published as British Standard DD ENV 28041:1993).

ISO 8253 Part 1; 1998. Acoustics – Audiometric test methods. Part 1: Basic pure tone air and bone conduction threshold audiometry (published as BS EN ISO 8253-1).

ISO 8662-4:1994. Hand-held portable power tools. Measurement of vibrations at the handle. Grinding machines (also published as BS EN ISO 8662-4:1995).

Kinnersley, P. *The hazards of work*. London: Pluto Press, 1973.

Lawson, I.J., McGeoch, K.L. A medical assessment process for a large volume of medico-legal compensation claims for hand-arm vibration syndrome. *Occupational Medicine* **53** (5), 2003.

Lines, J.A., Stayner, R.M. Ride vibration: Reduction of shocks arising from overtravel of seat suspensions. HSE Contract research report 240. Sudbury: HSE Books, 2000.

Ministry of Labour. *Noise and the worker*. London: HMSO, 1963.

Paddan, G.S., Griffin, M.J. Use of seating to control exposures to whole-body vibration. HSE Contract research report 335. Sudbury: HSE Books, 2001.

Paddan, G.S., Haward, B.M., Griffin, M.J., Palmer, K.T. Hand-transmitted vibration; Evaluation of some common sources of exposure in Great Bratain. Contract research report number 234. Sudbury: HSE Books, 1999.

Paddan, G.S., Haward, B.M., Griffin, M.J., Palmer, K.T. Whole-body vibration; Evaluation of some common sources of exposure in Great Britain. Contract research report number 235. Sudbury: HSE Books, 1999.

Palmer, K.T., Coggon, D., Bendall, H.E., Pannett, B., Griffin, M.J., Haward, B.M. Hand-transmitted vibration; Occupational exposures and their health effects in Great Britain. Contract research report number 232. Sudbury: HSE Books, 1999a.

Palmer, K.T., Coggon, D., Bendall, H.E., Pannett, B., Griffin, M.J., Haward, B.M. Whole-body vibration; Occupational exposures and their health effects in Great Britain. Contract research report number 233. Sudbury: HSE Books, 1999b.

Pelmear, P.L., Wasserman, D.E. *Hand–arm vibration*, second edition, Beverly Farms, MA: OEM Press, 1998.

The Supply of Machinery (Safety) Regulations. Statutory Instrument, 1992 number 3073.

The Noise at Work Regulations, 1989. Statutory Instrument 1989 number 1790.

Index